PODCAST AUDIO

Make Your Show Sound As Good As Your Content

Barry R. Hill

Rivershore Press

Published in 2020 by Rivershore Press. Visit us at www.rivershorecreative.com.

ISBN: 978-1-7321210-0-3

The audio examples are available at www.rivershorecreative.com.

Cover design by Scott Cole at Churchill Strategies.

CONTENTS

WHY THIS BOOK?

The way I see things, there are three problems in this world: mean people, dopey drivers, and awful sounding podcasts. Not that this is a new phenomenon, of course (speaking now of audio). There's an old quote from the Syd-Aud-Con audio training organization—*If bad sound were fatal, audio would be the leading cause of death.* Well, in that case I've met my Maker hundreds of times on the way to work listening to bad-sounding shows. Don't take it personally—all of you enthusiastic artists are experts in things *other* than audio. And that's fine, but let's see if we can ease the pain a bit and help each other out.

Podcasting got its start way back in the early internet days, when digital devices allowed us to download web-based radio shows and other content for portable playback. The "official" idea of a podcast comes from the idea that common folks like you and me can download or even create our own show and "broadcast" it on digital music players such as...you guessed it...the iPod. It's largely been a cottage industry since the mid-2000s, growing, but still dominated by amateurs who just want to be creative and have an outlet for their ideas. Now, however, it's rapidly transforming into serious business and becoming its own thing in the entertainment world. Of course there are still zillions of us cranking out shows from our bedrooms and basements, and that's why this book came about.

The best sounding shows, generally speaking, are high-level productions from the various NPR affiliates and other professional operations. That's our goal, even if we don't have $3000 microphones and million dollar studios. Possible? We can certainly move the needle a few clicks, so let's get started. This book is not intended to make you an audio engineer, but understanding some basic concepts and good practices will go a long way. And yes, you might have to spend a few dollars, but if you're serious about improving your shows you have no choice. Don't worry, you won't have to sell your firstborn or that brand new two-seater that'll stay in the garage half the year. Just let me drive it once in awhile.

We'll start by walking through a simple one-track recording, showing the equipment, procedures, and how to make it sound good. Then we'll break this down in more detail, saving the more complicated theory and such for later if you're so inclined. The website has audio examples so you can hear what we're talking about (it is audio, after all), so find a really good pair of headphones. Earbuds and those little speakers that came with your PC won't cut it, although they're great for comparing how your shows will sound in the real world. Of course you need your own gear, even if it's just free software and a single microphone. Experiment and follow along as much as you can so the reading makes more sense. It takes practice to get good at this, so let's get on with it.

ONE
FROM MICROPHONE TO MP3

We're going to walk through each step for making a basic recording. Along the way we'll briefly discuss the equipment you need, procedures to follow, and how to make it sound decent. You'll then have a pretty solid understanding of what's involved so we can dig a bit deeper as the book goes along.

Recording a track

Recording a source involves a microphone, an audio interface or microphone preamplifier, and a recording device. The goal is to select a mic that works well for that particular source, put it in a spot that sounds best, and get a good signal into the recorder.

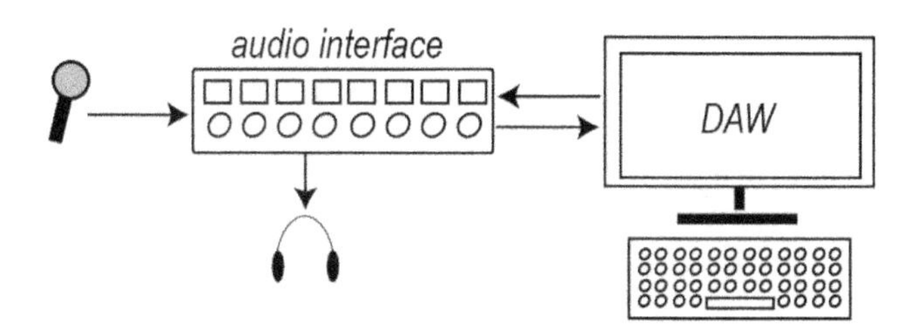

Equipment needed

- Microphone
- Mic stand
- Mic cable
- Pop filter
- Audio interface or microphone preamplifier
- Recorder/software
- Headphones/monitors

Microphone

If you have a USB mic, go ahead and connect it to your computer. Otherwise grab a standard mic cable and connect it between the microphone and audio interface, making sure you hear a click that indicates it's locked in. Set up the mic about six to eight inches from your face, slightly off to one side. If you have a pop filter, position that in front of the mic without it touching. The filter reduces those annoying pops and booms that come from talking directly into the mic. The off-angle placement also helps with this.

Audio example 1: Mic pop

Make sure you have the front of the mic facing you. Sometimes this is obvious, other times not so much. For example, long thin mics like the Shure SM57 are pointed with the end of the grill facing the source. Large diaphragm mics usually pick up from the side of the grill; one of these sides is the front—look for the manufacturer's nameplate or use your ears to make sure. When a sound is picked up away from the on-axis direction of the microphone, it will have an unnatural or distant sound quality, which generally doesn't sound very good.

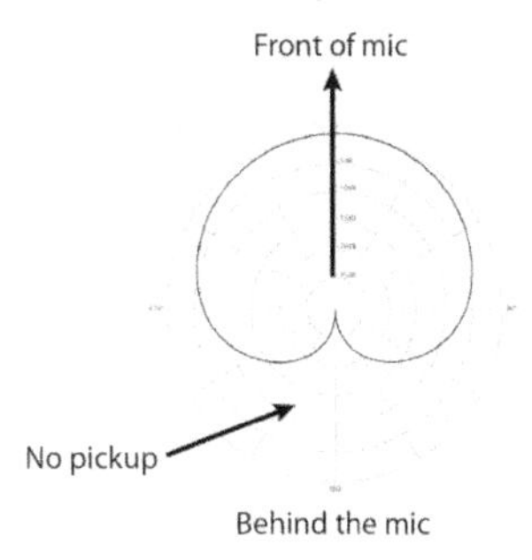

Check the switches, if any, on the mic. If yours has switches, it will be one or more of the following:

- *Polar pattern select*: You generally want the mic to pick up sounds from the front only, not from the back or sides. Set the polar pattern switch to the heart-shaped symbol; this represents a cardioid pattern and means the mic will pick up sounds mostly from the front.
- *Attenuation pad*: This prevents a really loud sound source from overloading (distorting) the microphone capsule. Your voice won't ever come close to this, so leave it off (odB).
- *Low-cut filter*: The symbol looks like a division sign; this attenuates (reduces) low frequency sounds such as rumble, vocal pops, and trucks driving by. This can also be done in software, the advantage being you can adjust exactly where this cutoff begins. That's where I'd do it, so leave this switch off (flat line).

Pay attention to *where* you're sitting. Room acoustics play a major role in what you're getting in the mic and is one of the prime factors contributing to our podcast quality plight. A plushly-furnished living room sounds much more controlled than a reverberant kitchen or bathroom (I'm sure it's been done). Spoken word requires a tightly-controlled acoustic environment, so find a smaller room with lots of drapes, couches, bookshelves, or acoustic panels on the walls. Make sure there's not a bare, undecorated wall close to you and the microphone; this causes direct reflections back into the mic and sounds terrible.

Audio example 2: Recording in a highly reverberant room

Audio example 3: Good room and mic placement

By the way, don't connect, disconnect, or move a microphone when the recording channel is on. It'll pop or make a loud noise through the system and into your headphones. Even if it doesn't damage anything it'll certainly not be fun on your ears. Disarm the track and turn the mic preamp down.

Audio interface / preamplifier

Some microphones are designed with a USB connection built-in, rather than a standard cable connector. This is really handy as you can plug it directly into the computer. These don't sound quite as good, though, but for podcast use they'll work fine. All other recording mics have the 3-pin connector we described earlier, called an *XLR*. The signal output of XLR microphones is very low and requires a special amplifier to increase it so it's ready for recording. This is called a *microphone preamplifier*, and it's always included in the audio interface or mixer.

An audio interface is the bridge between microphones and computers. Plug the mic cable into the XLR input on the interface, then USB from the device to the computer. The interface handles the analog to digital conversion for recording, then reverses this when playing back through headphones or speakers connected to the device's monitor outputs. This is a fairly complex, crucial internal operation, and as such you generally get what you pay for. Cheap interfaces will sound, well, cheap, so as the budget allows consider investing in something decent.

Some audio interfaces feature a single microphone input, some two, and so on. If you're planning to record multiple mics at the same time, such as for a group discussion, you'll need an interface with several mic inputs.

Typical settings on the interface to look for:

- *Headphone/monitor volume*
- *Input level*: sets incoming microphone level for recording. Leave this all the way down for now.
- *+48V*: Phantom power on/off switch. This is required for condenser mics (but won't hurt others if you're unsure).
- *Pad*: This switch will attenuate incoming signal, such as when the source itself is too loud for the input even when turned down all the way. Leave it off unless you're planning to beat a snare drum an inch away from the mic.
- *Low-cut filter*: Leave it off and use the ones in your software.
- *Source/mix balance control*: Set to mix so we're hearing only the output from the recorder.

Recorder / software

This could be a software DAW (digital audio workstation such as Pro Tools, Logic Pro, or Audition) or a portable flash recorder. Open a new session or file and set as follows:

- Bit rate: 24 bit
- Sample rate: 44.1kHz
- BWAV or WAV

For a DAW, look in the audio settings menu for how to select the audio interface hardware you're using. This will tell the software what inputs and outputs are available on the device as well as how to play back your tracks through monitors or headphones.

If it's an empty session, create a new audio track. Set the track's input to the audio interface channel where your mic is connected. For example, a stereo interface will have two mic inputs, so assign track #1 in the DAW to mic input #1.

Arm the track, meaning set it to prepare for recording; it should flash or turn red somewhere. Start speaking into the mic and gradually turn up the input level on the interface. The DAW track fader won't have anything to do with recording level, so just ignore it. Watch the meter on the track and set the input level so you have a healthy signal that peaks in the upper region, say no higher than -6dB, but well away from the very top. Hitting zero in digital, which is as far as it will go, will turn into instant hash as the system runs out of bits to encode the signal.

Audio example 4: Preamplifier distortion

That's it—press record and do a take. Keep an eye on the meter; it'll likely be somewhat higher once you get into the groove of your copy. Don't do any abrupt level changes, but if it's peaking out you'll have to do another take with a lower level (turn down the interface mic gain control).

If you want another go at it, return the DAW transport to the beginning of the session and just hit record again. It'll save each take you record; look for the clip list window to see everything that's created in the session. Make sure you name each audio track as soon as you create it. As you compile multiple takes and do editing on a track the clip list will automatically use that track name, which is hugely helpful. For example, would you rather have a list of tracks named Audio 1_1, Audio 1_2, Audio 2_1, Audio 2_2, and so on, or Freddy_1, Freddy_2, Mel_1, and Mel_2? Another big tip is to drop markers, or memory locations, along the way, such as when you said something that ought to be bleeped or you dropped your face into the mic. Then you can quickly go back and find these spots during editing.

Editing & mixing

Here are the steps to take our recording and produce a final mp3:

- Editing
- Mixing & signal processing
- Exporting a final file

Editing

For our current super-simple show example, editing should be as simple as getting rid of extra space at the beginning and end of the take. Unlike the old days when we took a razor blade and carved up the tape, editing in a DAW doesn't change the actual audio that was recorded; every edit, fade, or track adjustment is captured by the software, *describing* what should happen when the final file is exported. This allows you to experiment and keep trying until it's just right.

Look at the beginning of the waveform and you'll see blank space before you started talking. Select the *trimmer* tool (or whatever your software calls it), then drag from the left edge of the waveform toward the right until it's close to where the audio starts. Apply a short fade-in just before the audio begins so you get a smooth start. Zoom in as necessary to make it easier to see and select exactly what you want. Now grab the entire audio track and slide it left to the very beginning of the timeline. Play the file and it will begin immediately.

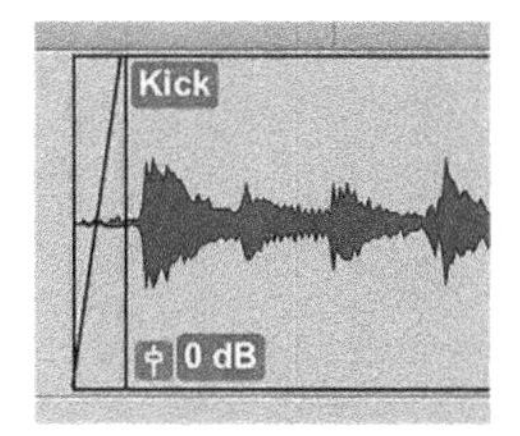

Quick fade at beginning for a smooth start

Pro Tools users should get familiar with *smart tools*, which means as you hover the cursor around the waveform different tools will appear. This is an incredibly efficient way to work: hover near the edge and the trimmer tool is selected, top half of the waveform gets the selector tool, bottom half is the grabber tool, and near each top corner a fade tool pops up. To enable smart tools, click the bracket over the set of main tools in the top window.

Smart tools enabled

Now trim the end of your take and apply a smooth fade out. By the way, I use Pro Tools for my audio work, but all of this translates to whatever software you have. You just have to find the equivalent tools and procedures in your setup.

Mixing and signal processing

Mixing a project means we're blending everything together, such as balancing the volume between show hosts and the music. In this case, though, it's only a single voice track to work on, so we'll concentrate on making it sound good. Recorded audio nearly always needs some type of processing to correct problems and/or make it sound better. We'll dig more into this later, but let's try a few basic things for the track you just recorded. Each of these processors are accessed via *plug-ins*, which we insert on each audio track (there will be a special placeholder for plug-ins, usually called *inserts*).

Filter/EQ

I always put on a low-cut (high-pass) filter. Filters attenuate (reduce) energy in a particular region. Unless you're Barry White, your voice does not extend all the way down into the lower frequency ranges, which is where we find miscellaneous noise, rumble, and so on. Click on a track insert and select an EQ, such as the 7-band EQ that comes with Pro Tools. Turn on a low-cut filter and adjust the frequency up the scale until it begins to cut out the lower portion of the sound. Backtrack a tad and you're set. If you're not sure, set it at 80Hz and you'll be fine. The other control, *slope*, determines how abruptly it will chop off this frequency range. For now, set it for a fairly steep slope (24dB/oct).

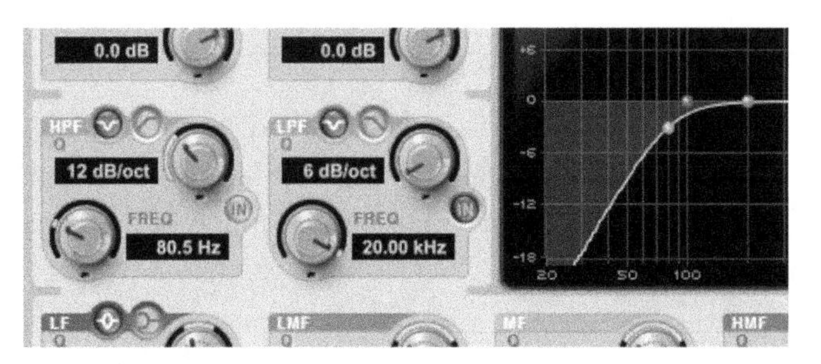

High pass filter turned on with controls on the left

Audio example 5: Low cut filter (bypassed, on)

The next step is to clean up the mud or cloudiness that's common on acoustically-recorded tracks. Find the low-mid band on the EQ and turn it up 6dB or so. Sweep the frequency select control back and forth between 300 and 500Hz and listen for a spot that sounds more muddy or cloudy than the rest. It'll all sound weird, and it'll take lots of practice and listening to get a feel for it. Once you zero in on this, turn the level control down to -3 or so. Bypass the EQ and listen to the before and after; it should be clearer with the EQ turned on. If it sounds too thin, turn the level back up a bit.

Audio example 6: Low-mid cloudiness, then reduced with EQ

This same approach can be used for an edgy voice; I had a show recording once where the host used a mic with a high-mid frequency boost switch inadvertently left on. It was way too bright and harsh, so I set the high-mid EQ band around 3kHz or so and turned it down a few dB.

Audio example 7: High-mid edginess, then smoothed with EQ

There's a third control on most EQs called *bandwidth* or Q. This controls how wide of a region the EQ will change. Keep this fairly narrow for this step so it doesn't pull out too much of the overall sound of the track.

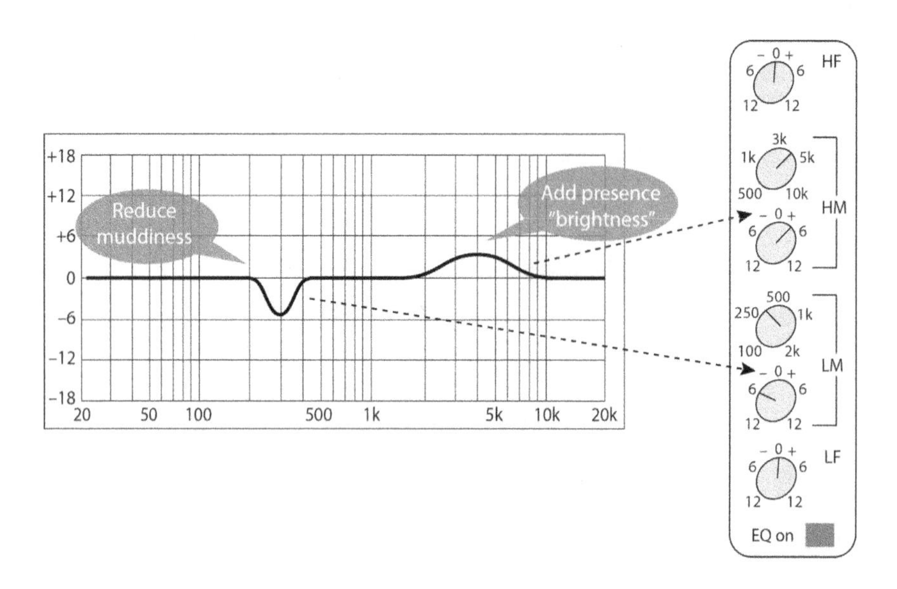

Now that we have a clean, basic sound, let's shape it a bit so it has more presence and sparkle. Grab the high-mid control and boost it 2dB. Set the Q for a wide bandwidth and sweep around the 4k region. This is a good area for presence and intelligibility. Don't overdo it, so perhaps a couple dB is all that's needed, depending on the source. Now try a very gentle boost with the high frequency control. Set it to shelving mode and sweep up around 8-10kHz. This applies a flat boost to the entire upper frequency range, providing more "air" and openness. Sometimes I'll just go with the high shelf boost without the high-mid band. Lastly, try boosting the low-frequency control a couple dB around 200-250Hz. Exact frequency regions will depend on the voice, so over time you'll get a better feel for how to treat different situations.

Here's a voice track from a show that starts with the EQ bypassed (no processing), then with EQ turned on. The graphic below shows the settings.

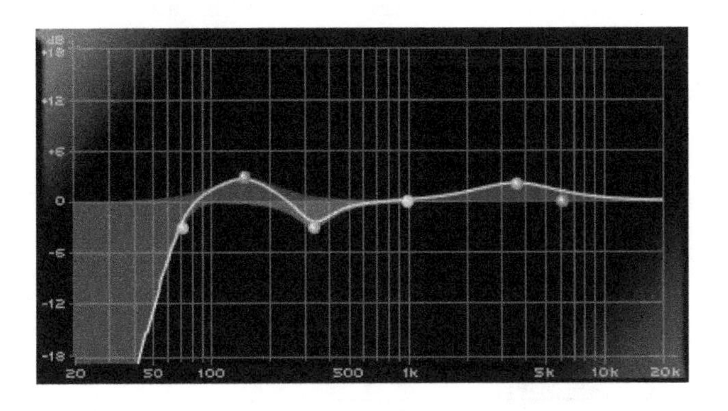

Audio example 8: Original voice track, then with EQ

Compression

Audio compression is different from data compression. In this case we're taking a wide dynamic range of a recording, meaning the swings from soft to loud, and reigning it in a bit so it's easier to listen to. This is especially important where most folks listen to your shows—on the subway, in the car, or mowing the yard. You don't want the softer passages or phrases getting lost in the surrounding noise, so we'll flatten this variability and increase our final signal level to maximize what the listener hears.

The first step for this is putting a compressor plug-in on the voice track. Set this after the EQ and set the ratio to 3:1, fairly fast attack, and a medium release. Now play the track and adjust the threshold down until you see a few dB of gain reduction on the meter. Lowering the threshold will compress more of the track's dynamic range whereas a high setting will only affect the peaks (louder portions). Once the threshold tells the compressor to start, the ratio determines how much it will reduce the output. So a 3:1 setting means that only 1dB will be output for every 3dB coming in over threshold. We don't want to squash this too much, just level it out a bit, so if the LEDs are lit up like a Christmas tree you might want to back off the threshold a little. As a nice side benefit, compressors also add a little beef to the sound, making it fuller.

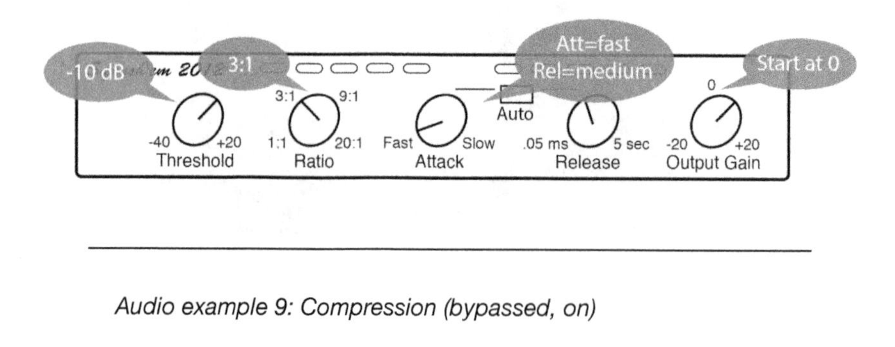

Audio example 9: Compression (bypassed, on)

Reverb

This one is simple. Don't.

Oh, you want to know why? We love reverb in music production, but for podcasting the goal is to have as clear and intelligible a voice recording as possible. Reverb destroys that, so we hate it. Start with your mic setup and reduce the amount of room reverberation you capture into the mic, and for sure don't add more later.

Final steps

Once you've got things sounding like you want, let's get ready to export it to a new audio file. We're going to add one more processor to our simple one-track recording that will make your show sound louder and more consistent. Insert a *peak limiter* after the EQ and compressor; peak limiters are overused these days, but the idea is to take lower-level audio and increase it while maintaining a maximum ceiling on the overall track. By squashing dynamic range into a smaller space, the final show file sounds louder and fuller; overdo it and it starts sounding like hash. It's like a compressor on steroids, but it's not the same animal; you can't just crank up your audio compressor to accomplish the same thing.

Be moderate with the settings. Pro Tools comes with a peak limiter called *Maxim* (*adaptive limiter* in Logic Pro); set the output peak level for -1.5 dB and slowly decrease the threshold while playing the track. Listen as it gets louder, then progressively more nasty; find a balance that sounds louder but still natural. There's a better method to all of this, but we'll cover that later; for now I wanted to focus on tools you probably already have.

Audio example 10: Peak limiter bypassed

Audio example 11: Peak limiter moderate setting

Audio example 12: Peak limiter overdone

Finally, make sure the overall signal level is not prone to distorting. As you skip through the recorded track at various points, especially the really loud ones where your cappuccino starts kicking in, see how high the track meter goes. Target a level range around -6dB, making sure any peaks (a red indicator should flash) are avoided by turning down the fader. Conversely, if the overall signal is way down, turn the fader up.

Now let's create the final show file. This is called *bouncing* or *exporting*, so go to the file > bounce menu and set things as follows:

- *Bounce source*: Main or Mon LR
- *File type*: WAV
- *Format*: Interleaved/stereo
- *Bit depth*: 24
- *Sample rate*: 44.1kHz

The offline option is non-realtime, meaning it will render the file much quicker (a good thing). Most podcast shows are streamed as mp3, so include this in the bounce if your software will do it. If not you can convert it later. Keep the WAV version of your final show in case you need to go back and redo an mp3; WAV files are far better than mp3 since they are uncompressed, just like a CD. Mp3 files are lower in quality, but feature small file sizes that are ideal for hosting and streaming; for now set the mp3 bit rate for 96kbit/s. (I run my show at 160, but we'll talk about that later.)

That's it—you've produced a complete recording from microphone to mp3. All of these steps will come up again in the following chapters, but with more detailed information, options, and recommendations.

TWO
RECORDING THE SHOW

Let's back up and take a deeper dive at some of the variables and procedures that affect your recording quality.

Microphone selection

Large diaphragm condenser mics are great for voice work. They provide clear articulation, sensitive pickup, and a full high and low frequency response. Condensers in general capture sound really well, even from a distance. This is both good and bad—you can move around a bit and not lose yourself, but they pick up lots more room echoes and noise, including the person sitting across the table from you. Many podcasters prefer dynamics for this very reason, and there are several models that are excellent for speech recording such as the venerable Shure SM7B. The trick for these is to stick fairly close to the mic; don't move around much or turn your head or the sound will noticeably change and diminish since dynamics aren't nearly as sensitive. But their real benefit is not capturing as much of the room sound around you, resulting in a cleaner track to process later. I prefer condensers as long as I can control the room decently; I prefer the richer, more detailed sound quality they offer.

There are loads of mics marketed as "podcasting mics", which means, well, often not much of anything except a USB connector and an affordable price tag. Some feature a built-in headphone jack and perhaps a mute switch. The switch is actually pretty handy for when you get choked up on your twinkies, but it can be added separately. Don't ever use the laptop mic or the cheap headset that came with your Dragon Dictation software; the quality for both is awful, and the laptop mic is too far away to avoid room echoes and noise.

Now, a good quality headset mic is attractive because the mic capsule remains steady in front of your face no matter where you're looking and moving. Most of these are designed for wireless applications, such as for a lecturer or pastor, but you can get a wired connection instead. Do this and avoid the headaches with wireless systems such as interference, dropouts, random noises, and batteries. Some of these hook over only one ear and can move around slightly; use a small piece of tape (yes, to the face) to nail it down. The two-ear models are more stable.

For a group of people talking around a table, the best scenario is for each person to have their own mic. But if that's not possible, use a single mic that picks up all around—not just in front of it. This is called an *omnidirectional* polar pattern, and some manufacturers emphasize this function by shaping it like a ball (totally unnecessary, but it looks cute). Some condenser microphones feature switchable patterns, so they can be set to omni for a situation like this or cardioid (uni-directional) for standard, in-your-face placement.

A basic $60 USB podcasting mic will do fine to get you started—there's no need to spend $800 on a studio quality mic. As you develop your skills, though, you'll begin to hear the difference and will probably want to step up to the next level. Price alone doesn't always dictate good or bad, so search around and find out what professional podcasters are using. I've tried a variety of mics for the co-hosts on our show and was surprised at some of the results. What I thought would be a no-brainer sounded terrible on certain people. Even for myself, I tried one of those awesome Shure SM7B mics, which are seen everywhere, and it was awful on my voice. Now, my voice is pretty bad regardless, but switching to my vintage large diaphragm condenser was night and day. Ask your local audio dealer if you can try a mic or two until you settle on a favorite.

A decent mic stand is important. Cheaper mics come with a tiny stand that places the mic too far away. Look for a guitar-amp stand or something similar that you can set on top of your desk, and if you're serious about doing this right, find an adjustable boom arm that mounts to the desk. These are great because the arm keeps things floating out of the way, placement can be precisely adjusted, and it helps isolate vibrations such as when you bump the table with your knee. It can also be pushed aside when not needed.

USB or XLR?

USB microphones are popular because they connect directly to the computer with no extra hardware needed. I have one that cost about $250 and sounds pretty good. Keep in mind there's certainly a difference between this and a $60 model, so invest the extra money if you can. If you're recording more than one mic at a time, you have to go with XLRs; you can't plug in multiple USB mics.

The other route is a standard XLR mic, so-called because of the 3-pin audio connector it uses to interface with professional audio equipment. This is an analog signal, so the mic is plugged into either a mixer or audio interface and must be converted to digital for recording and processing in software. These devices provide a *microphone preamplifier*, which is required to take the very low-level signal coming from the microphone and increase it to match what the mixer or recorder need to operate. Phantom power is also supplied for powering condenser mics; make sure the switch is on unless you're using a dynamic (it won't hurt a dynamic if left on). The other important function is the actual conversion from analog to digital, which directly influences the sound you get. Along with the choice of mic, placement technique, and room acoustics, the audio interface is key for sound quality. You get what you pay for, so a more expensive model will have high quality preamps, premium analog-to-digital conversion, plenty of gain, low noise, and should last forever. One mic preamplifier input is required for every XLR mic used; some interfaces come with one, some two, and others might have up to eight.

Of course, there are cheap XLR mics that cost $50 and, on the other end, amazing models that cost thousands of dollars. Although I like the USB mic I mentioned earlier, I also have a classic Neumann that blows most anything out of the water...but it runs $3600. And I plug it into a $700 interface. Yeah, you could say it sounds pretty good. But money doesn't solve all things, however. As I mentioned, I tried the ever-popular Shure SM7B, which runs a moderate $350, and hated it on my voice. Microphones are not created equal for reasons besides just cost, so you'll have to experiment until you find that magical moment when you've found your mate. Just don't skimp on the preamp/converter/interface, which does generally follow the money trail.

And don't forget the really expensive mic in your pocket. iPhones aren't so bad if placed in a good spot. It might not be a Neumann, but it'll get the job done. Just make sure you turn it around so the mic is facing the source.

Audio example 13: Neumann TLM103 condenser

Audio example 14: Audio-Technica AT2020 USB condenser

Audio example 15: Shure Beta 87A condenser

Audio example 16: Telefunken M82 dynamic

Audio example 17: Shure SM7B dynamic

Audio example 18: Electro-Voice RE20 dynamic

Audio example 19: Laptop mic

Monitoring

Headphones are essential for monitoring what your voice sounds like as well as hearing anybody else you're recording with. If it's a Skype or remote recording, you can't use speakers in the room because your mic will pick that up. Even if it's just you or you and your co-host in the same room, using headphones will enable you to hear exactly what your voice is sounding like. Doesn't seem like a big deal, but we often don't realize we're moving around and making weird noises. Headphone monitoring will give you immediate feedback (the good kind) so you can correct and move on rather than discover an ongoing problem when you're mastering the final show file. Now, does this mean you have to force all your guests sitting around the table to wear them? Nah. Don't make them uncomfortable. As long as someone is monitoring the recording you'll be okay. Unless of course the conversation includes someone coming in from Skype. Then it gets more complicated.

Inexpensive earbuds won't cut it—you need something that will isolate sound around you so the mic won't pick up what's coming through the phones. High quality isolating earphones can work pretty well; these are the type musicians will splurge for when they play live shows. Most podcasters use regular headphones, but they have to be good, and they must cover the entire ear. And the challenge is that all headphones are not created equal. Sure, there are cheap ones and really expensive ones, but even the professional models all sound different. I have three sets, and none of them are perfect for what I'd like to hear. So I have to get used to one of them and understand how to compensate accordingly. How's this? I've used the professional Sony MDR-7506 for decades. They're great, and they only cost about a hundred bucks. You see them everywhere. They're just a bit bright in the high frequency range, however, so I keep that in mind when setting EQs and such. The trick is, whatever pair you end up with, listen to a lot of good quality shows for comparison. This gives you a reference to model your work by.

How about speakers, at least during mixing? Sure, some podcasters have a nice monitor setup for producing their shows. But they have to be good ones with full-range frequency response. This means none of those small PC speakers and preferably honest-to-goodness *studio monitors*. These are designed for accurate reproduction, not impressing your friends with enough bass to bounce the goldfish out of the bowl. Whichever route you go, headphones or monitor speakers, follow the same advice I just gave—listen to lots of great audio, get used to how they sound, and tailor your shows to match that as best you can. One more bit of advice? Headphones are far cheaper...just sayin'...

Recorders / software

Recording shows in your home or office setup should be done with an audio interface and computer/iPad with recording software. You can keep a better eye on recording levels and don't need to transfer any files from a portable recorder. For recording in the field, portable hand-held recorders are very convenient. They're small, record to flash media such as SD cards, have built-in mics, and can be mounted on a standard camera tripod. Some models have preamps for connecting external mics to get better audio quality. You can also take your laptop, audio interface, and studio mic setup. This is more cumbersome, but results in a higher quality recording, all things being equal. Too much stuff? Take your iPad or iPhone along with your good mics; there are recording apps and audio interface options for iOS devices.

With podcasting fever spreading so rapidly these days, manufacturers are taking note and designing audio gear purpose-built for producing shows. As of this writing, we're beginning to see combination mixer/recorder units with multiple mic preamps, multiple headphone jacks with individual level controls, assignable pads for triggering sound effects and music cues, and even phone connections for recording remote guests. They record to either SD cards, flash media, or even direct to your computer via USB. All of this is super convenient; after the recording session simply edit and mix in your computer software.

All modern recording equipment will record audio at high resolutions, meaning a minimum of 44.1kHz (CD quality) and 24-bits. Read over the digital audio chapter for a technical explanation of what all this means, but always set your equipment at this setting or better. Never record directly to mp3 or AAC, which are highly compressed files; you want the best source material possible that can be processed and adapted as necessary later. It's the same as snapping a photo with a low-quality jpeg setting and wondering why it's so pixelated when printing to poster-sized stock.

Watch the meter and set the incoming mic levels so they're not close to hitting zero (top of the scale). 0dBFS (full scale) is the absolute maximum for a digital recording system; anything beyond that turns into hash as the system runs out of bits to encode the signal. I can't emphasize this enough. You and any other guests on your show are having a spirited, fun conversation (or something better than boring, let's hope). This means a wide dynamic range, with bursts of laughter, shouting, whatever. Don't push close to the top. With 24-bit recordings we can keep a healthy safety margin without losing quality. Though you don't want to record too low, it's not necessary to push the top limit like we did in the old days. Current technology has more than sufficient dynamic range (quiet to loud), so just get it safely in the upper-middle range and you'll be in good shape. Distorted recordings sound absolutely terrible and can possibly be repaired to some extent only with expensive software (and time).

Audio example 20: Overloading the preamp

Speaking of software, lots of podcasters start out with Audacity. Why? It's free, which is a pretty good incentive. Once you get a quality conversion from the microphone through your interface, it's somewhat less critical what application you use for recording and processing. However, you certainly want something that handles multiple tracks so you can control each host, guest, music bed, and sound effect separately. *Audio editors* only record mono or stereo files and do not allow the editing flexibility like you have with digital audio workstations. A DAW provides individual tracks for each part, complete freedom to edit and move things around as you want, built-in processing such as EQ and compression, and exporting of a mixed file.

GarageBand can get you started and is also free (once you buy the Macintosh); if you happen to have the Adobe Creative Suite it comes with Audition, which is far better than GarageBand. There are several options at all price points, even some that do everything online. Ferrite is popular for the iPad, especially for podcasters on the go who are editing shows in hotel rooms. A couple of manufacturers bundle their audio interfaces with a DAW; though it might be a version with limited features, it should be more than plenty for what you need. That's a better route than Audacity for sure, which doesn't sport the most beautiful interface and doesn't quite work like a DAW. Once you become more sophisticated and have advanced needs for editing and processing you might step up to something like Pro Tools or Logic Pro. Both of these are very powerful and come bundled with most of the processing options you need.

To summarize the signal flow for recording shows on your computer, the signal chain goes from microphone to audio interface, which connects to the computer via USB, then back out to the interface for monitoring through headphones.

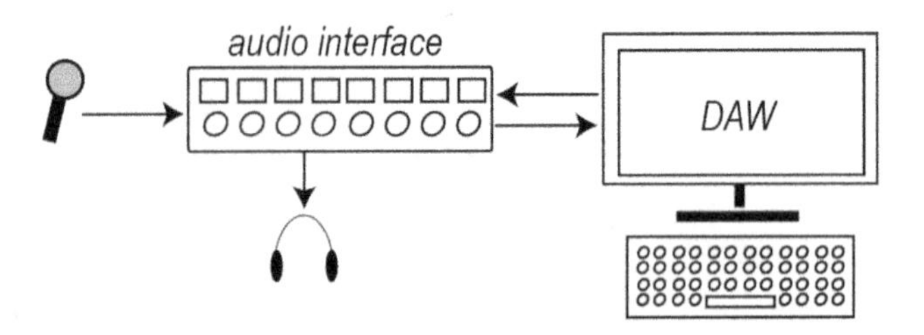

Having problems getting signal? Try these steps:

- Mic cable plugged in all the way on both ends?
- Mic attenuation pad turned off?
- Does the mic itself have an "on" switch?

- Or mute switch?
- Phantom power on (for condensers)?
- Audio interface preamp level up?
- DAW track set to the correct mic input?
- DAW track in record mode?
- Try a different mic cable.
- Try a different mic.
- Try a different Snapple flavor.

Multi-location recording

Many podcasts are recorded with participants scattered all over the place. Usually these are phoned in or connected via VoIP services such as Skype. To record this we need to connect either a phone (for a voice call) or phone, iPad, or computer for a Skype conversation. There are a variety of methods, software, and hardware for achieving this, some of it straightforward, some not. Essentially our options are:

- *Overall Skype recording*: Use the built-in Skype recording function to capture the entire conversation.
- *Double-ender*: Each participant records their own voice using a mic at their location.
- *Separate Skype track recording*: Record the remote guest, either from phone or Skype, onto a separate track from your mic. This approach requires certain software, hardware and/or a more complicated setup.

Blended Skype recording

The simplest way is to use Skype itself to capture the conversation; the main drawback is that it sounds like, well, a Skype conversation. It also blends everyone on the call together, so you can't treat each voice separately when mixing the show.

Double-ender

The far better method is what's known as the "double-ender", meaning each participant records their own voice track and sends you the file afterwards. If each remote co-host takes time and a few dollars to build a decent setup at home, the show will sound exponentially better. Guests of the show, however, typically aren't set up to do this, so you have to make the decision whether to pursue it or just record the Skype signal. It doesn't have to be complicated—they can use their computer's built-in audio recording application, even if the only mic available is the built-in one. Tell them to select the microphone as the input device, choose WAV and 24-bit, then press record. Coach them on how to sit relative to their mic or computer, to use headphones in order to avoid bleed, and how to set their recording levels so you get something usable.

After the session, have your guests send their audio recording to you and import into your software. One trick for syncing everything in your DAW during editing is to use a master recording from your Skype session, which includes everything, as a timing reference. Import the individual tracks from your participants and visually match their waveforms while listening to the voices. You can also try having everyone clap on a countdown at the beginning, but this is hardly exact; ultimately you'll adjust here and there as necessary for everything to feel right, especially for longer shows (it'll drift over time).

If Skype connections are driving you crazy, try a web-based service such as Source-Connect Now. Everyone in the conversation participates via an easy-to-use browser interface that provides much higher quality audio than Skype. For now it offers the same limitations: each person needs to record their own mic audio and send you the file; otherwise you're stuck with the overall web audio stream (which at least sounds better than Skype). Even though their system provides a recording function (to record each person's local mic), it only captures 16 minutes at a time—not usable for what we need. The basic service is free, with add-on premium features promised down the road (unlimited high quality recording of each user would be awesome). If you're a Pro Tools user, you could consider installing the company's $125 Source-Nexus AAX plug-in, which records directly to your session timeline. But I'd probably stick with another application called Audio Hijack (keep reading below); I like the simplicity and nearly unlimited flexibility it provides.

Separate Skype track recording

Sometimes it's not reasonable to have your guest record their side of the conversation; you have to get it done on your end, but we also don't want to record a Skype audio signal that has everything blended into a single track. This requires just a little more thinking with signal flow and perhaps an inexpensive piece of software. The idea is to record your local mic while sending that signal to your guest; you grab their incoming Skype audio and record that to a separate track while also feeding it to your headphones. This can be complicated or not-quite-so-much, so let's lean toward the not-quite-so-much.

Audio Hijack is an inexpensive software application that can grab any audio on your Mac and do stuff with it such as record to a WAV file, route it to another device, or even add processing along the way such as EQ or compression. In this case we want to work with separate audio sources, such as your incoming Skype guest and your local microphone. The example below is a very simple configuration with the incoming Skype audio (your guest) being sent to your interface so you can hear it while also recording it as a WAV file. Your local mic is also being recorded as a WAV file. Both of these audio files would then be imported into your DAW for editing and mastering. Note that with multiple Skype callers it cannot separate those into individual audio files, so you would need to go with the double-ender.

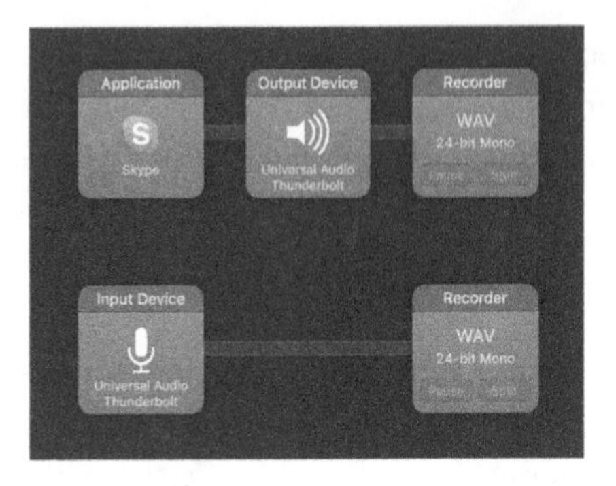

Do you already have a USB hardware mixer? If so, here's another (more complicated) way to accomplish this without buying more stuff; just connect the mixer to a DAW on the computer. Remember the goal is to provide local microphone audio to the remote guest *without* including their Skype feed, which would create a delay on their end (they hear themselves speak in the room, then their mic audio comes back a split second later after it's run across the web and back). The typical solution is known as a *mix-minus*, where you send an individualized feed that avoids the delay. Try this on for size:

- USB mixer
- Local mic(s) and headphones
- Computer running the Skype connection

Plug your mic into channel one of the mixer, then connect the output of the computer (Skype audio) into channel two (1/4" line input). The DAW will now record both sources (or more, if you've got 'em) as separate tracks. Plug your headphones into the mixer so you can hear everything. The trick now is to get everything *except* the remote Skype audio back to that guest so they can hear on their end without the annoying delay—hang on for the ride:

Connect the mixer's Aux 1 send (output) to the computer's audio input (set that as your Skype microphone/input source), then turn up the Aux 1 control on all active mic channels *except* for the Skype channel. Essentially the aux send is a sub-mixer, creating a separate blend that you can customize for the remote guest. If they need more of local person #2, then turn up that mic's aux send.

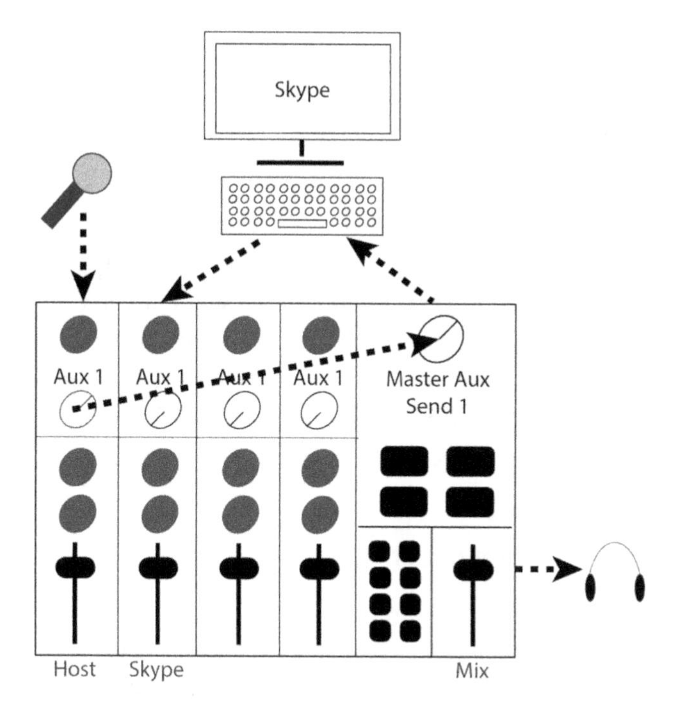

By the way, if you have multiple people in the room with you, they each require a headphone feed so they can hear the remote guest. Simply plug the headphone output from the mixer into an inexpensive headphone splitter, which provides multiple outputs with corresponding volume controls.

You can do this with an iPad, but you'll need an iOS audio interface to make the two-way connection with the lightning jack (or whatever they have on them by the time you read this). Computer cabling can also be tricky, because you're dealing with a stereo input and output on the computer, connecting to mono channels on the mixer. The computer will probably be 1/8" stereo output (3.5mm), and this needs to be split out into left and right 1/4"; connect the left 1/4" into the mixer input. Reverse this for routing the mixer output back to the computer. And of course many Apple computers don't even have analog audio input or output jacks.

Yikes. Is there an easier mix-minus solution, especially for when you've got multiple participants in the same room and Skyping with a remote guest? Buy a podcast-specific mixer/recorder, such as from Zoom or Rode. They have a built-in TRSS jack (3.5mm connector with three black rings on it) that you connect to your phone; the mixer automatically generates a mix-minus signal back to the phone. Genius. And for only a few hundred dollars.

Confused? Get help, call 911, or just hit record in Skype. The double-ender, though, is your best bet. If your guest(s) can capture a decent recording on their end it'll sound far better. Again, if you're bringing in multiple remote participants you have no choice but go for a double-ender or just use the combined built-in Skype recording. The good news is manufacturers will continue offering improved features for podcasters, so things can only get better, right?

Microphone placement & room acoustics

Closer is better for voice recording. Avoid putting the mic a few feet or more away from a person, especially if it's you as the main host. The goal is clear articulation, full frequency response, and minimal interference from the surrounding acoustics. The challenge is that microphones don't behave like people when listening in a room. If someone is talking across the room from you, it's pretty easy to focus in and hear what they're saying. We don't conscientiously register what the room is doing to the sound as it travels across.

Mics, on the other hand, get what they get with no bias or judgement capability whatsoever. If a source is very close to the mic, it will capture mostly that sound with little of the surrounding room environment. But move the mic just a little farther away and things begin to change.

When a sound is generated in a room it spreads throughout the space until it hits a boundary, such as a wall, ceiling, filing cabinet, or window. It then bounces back until it hits something else. Eventually the energy dies away, depending on how reflective or absorptive the room is. This is the difference between a hard-tiled bathroom and a living room. The comfy couch, carpet, plants, and curtains absorb sound energy and reduce reflections. This results in a drier, more controlled environment, which is ideal for voice recording.

Sitting in a kitchen with a laptop a couple feet away captures lots of room reflections, making the voice sound less clear, more reverberant. The "echoey" room sound will reduce intelligibility and make your show sound unprofessional. Even with a quality mic, if it's two to three feet away you'll get too much room sound. Another issue for distant mic placement is that sound energy attenuates (diminishes) over time and space, so the mic is getting a quieter sound. This has to be amplified by turning up the preamp more, increasing noise and the surrounding room sound. Finally, every room has background noises you've tuned out long ago...air conditioning vents, refrigerator hum, cars driving by. Condenser mics get all of this, but less so if the mic is pretty close to the main source. So proximity is key.

Audio example 21: Good mic placement

Audio example 22: Distant mic placement

Audio example 23: Laptop mic in a poor room

Place the mic a few inches from your mouth, angled just a bit over to one side to avoid direct "pops". What's a pop? Well, first check to make sure nobody's looking. Now hold your hand in front of your mouth and say words with "p" and "b" consonants. You can feel the puffs of air, especially downward from the mouth. This energy smacks the mic's diaphragm, causing a low-frequency pop in the sound. A pop-filter, usually a round screen made of nylon or metal, is helpful to reduce these pops as well. Mount this just in front of the mic without touching. When talking, stay pretty close to this position; a change of only a couple inches or so will make a difference in the sound, so try not to get too excited and dance around.

Audio example 24: Mic pop

Also notice that the closer you get to the mic, the more bass you'll hear in your voice. This is called *proximity effect*, and radio DJs have happily worked this effect for decades. If this floats your boat, just make sure you're not too close and getting pops into the mic. That's another reason for using headphones while you record—it's always better to know exactly how everything sounds now before you've recorded your two-hour epic episode.

For a group conversation, try to have one mic for each person. If this isn't possible, an omni mic sitting in the middle will do a halfway decent job picking up everybody as long as they're relatively equidistant from the mic and not too far away. You'll get more room sound, so it's not as "up-close and personal" as a regular mic placement. Sometimes you have no choice, though. It's especially important to treat the room itself so as to reduce reflections.

Fixing your room

Most professional voice recording is performed in an acoustically treated sound booth. This usually means there are absorptive panels on the walls and perhaps on the ceiling as well. The room is also fairly small, so it can't generate lots of room reflections and reverberation. Even if you don't have access to such a facility, there are fairly simple solutions for improving any space you're stuck with.

If this is a room dedicated for recording, it makes sense to invest in some acoustic sound panels for the walls. They're not expensive and can be ordered in a variety of fabric colors. If you're handy with your hands you could make these yourself. It's not necessary or advisable to completely cover the walls from one end to the other. Flat, bare wall surfaces are the enemy, so just spread the panels around the room to minimize large sections of untreated area. Other options are to hang curtains (thicker is better), install artwork that has depth and various dimensions, fill bookshelves with books of various sizes, or even roll a coatrack over with your winter collection. Put carpet or a throw rug on the floor. In general, soft and fuzzy stuff is best, so look for anything like this to set up around you.

Avoid sitting close to a wall; not only can these reflections make the voice sound reverberant, they can also destructively affect the quality, or tone, of the voice. What happens is that the sound waves of your voice go directly to the microphone, but also toward any nearby wall. Depending on the angles involved they'll reflect back toward the microphone. This is the same sound, but delayed in time because it takes longer to reach the nearby wall and over to the mic. If you remember anything about sine waves from school, take two copies of an identical signal. Now slide one over a bit in time and add the two together. As the various positive and negative points line up they'll construct a signal that doesn't look, or sound, like what you started with. Usually the result is a hollow, ethereal sound that lacks fullness and body. This is what happens when a nearby wall, filing cabinet, or other hard surface reflects sound back into the mic. A simple acoustic panel will do the trick, and again, staying closer to the mic will reduce the impact.

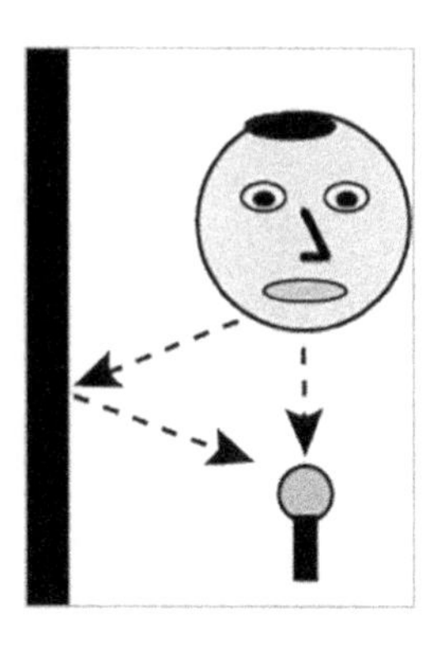

By the way, the same thing happens when you're holding a book or sheet of paper while you talk into the mic. Be careful about where these are positioned and listen carefully for any tonal changes that are occurring.

Outside noises or other stuff going on in the building where you're located is more difficult to deal with. Isolating sounds, meaning keeping them out, is far more difficult than controlling the interior acoustics of a room. At the very least record your shows in a room with a closed door as far away from major noise sources as possible. If there's a gap under the door, stuff it with a towel or blanket. Air gaps are the enemy, and doors are the worst offenders. Sit with the front of the mic facing away from the door to minimize shouting matches from the hallway. Schedule your sessions to avoid major issues, such as the daily ten o'clock Special running exclusive passenger service between Chicago and Nashville right past your parking lot. Beyond this, we're talking about building one of those fancy-smancy recording studios, so that's a topic for a different book.

Audio example 25: Phasing from holding a book close

Ok, go record your show and then come right back so we can put it all together. I'll grab some coffee (and a nap).

THREE
EDITING, MIXING AND MASTERING

Done? Let's take a look and get the show file ready.

Editing

Editing is necessary to get rid of dead space before and after the recording, add smooth fades to get in and out of the show, compile various segments, add an ad somewhere in the middle, and perhaps deal with unintended starts, stops, and stutters along the way. This can be quickly accomplished in your DAW; go look at the first chapter for basic editing technique.

Many shows are pieced together, such as with an introduction, the main content presentation, and a wrap-up. If these were recorded at different times and/or places, put them on separate tracks so each can be processed individually, preferably attempting to make them sound similar. You certainly don't want to string them all out on the same track and suffer dramatic changes in tone, quality, and level (volume). Music and sound effects should also go on their own tracks as well so you can EQ, pan, and set appropriate mix levels.

Each part in the show gets its own track

If you're inserting an advertisement into the show, try to find a natural break point. Some shows have it at the beginning, others at the end, still others break into the conversation in a feeble attempt to make sure listeners hear it. Look for a graceful way to fade out of the discussion, bring in the ad, then fade back in. A change of topics is best, and experiment with the exact placement of the break and fades to make it as smooth a transition as possible. Avoid abrupt cut offs and drop outs (and ins).

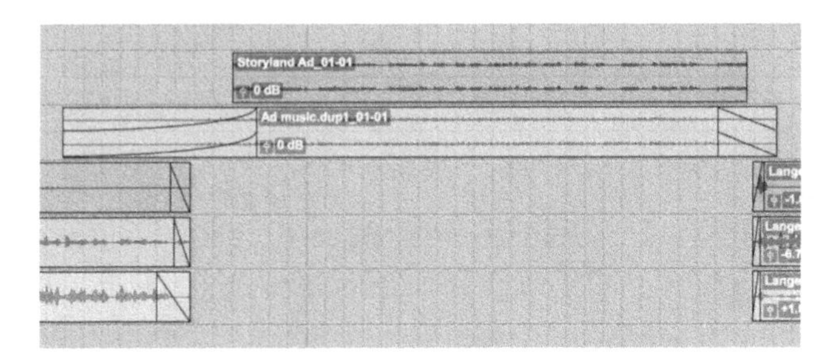

Note the gradual fade in for the music and how elements overlap for seamless transitions

---

Audio example 26: Smooth ad insert

Audio example 27: Abrupt ad transition with awkward timing

Now let's say the grandfather clock kicked in while you were gearing up for the next sentence, forcing everyone to pause for a couple moments (it's only two in the afternoon, thankfully). You can select all your tracks and edit that chunk out, but then what? Option A is to slide everything over toward the left to join the previous segment. If this works, apply a cross fade, listen through the edit to make sure it's smooth, and that's it.

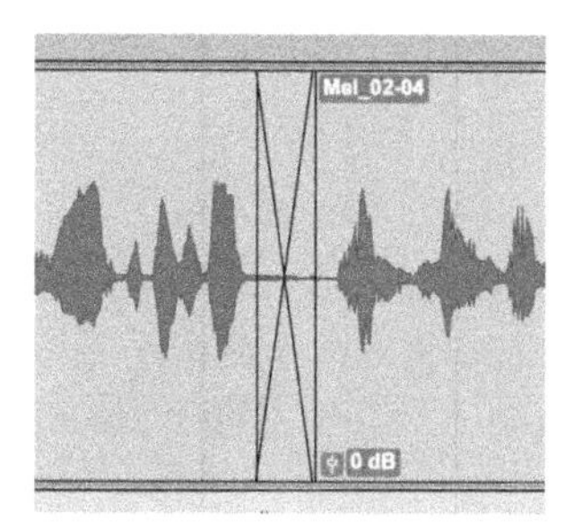

Crossfade to smooth the edit point

But, maybe now the next segment comes too soon and sounds rushed and unnatural. You need to leave a bit of space between them, which would leave a short section of complete silence. That's going to be obvious, because all of the recorded tracks have at least some small amount of what's known as "room tone"; there's always environmental noise, whether it's the AC running in the background or air movement in the room. The way they handle this in television and film production is to record several seconds of room tone; set everything up in the room, but without people present (or have them stay silent). Then you can grab a chunk of this to insert into that blank spot. Cross-fade on either side to smooth the joints.

If all the participants were located in different rooms and the clock only affected your recording, then you don't need all of this. Just edit out the clock chimes on your track; the other tracks will mask the resulting silence.

As you're editing and moving things around, remember it's all about timing as well—not time as in duration, but rather a subtle feel for how things should flow. Don't just stick them back to back on the track and call it a day. Nudge each section back and forth in small increments to make it flow, like an actor or comedian would do with their lines. It *is* show business, after all.

Audio example 28: Timing, first awkward, then better

Mixing

Mixing is the process of blending everything together into a final show. This involves processing (EQ, compression, etc), panning (stereo placement), and volume balance (faders). We want all the voices, music, and sound effects to contribute toward a smooth, seamless listening experience. To do all this each part needs to be on its own track, using the individual plug-in inserts, pans, and faders.

Processing

Each recorded track will need some amount of EQ and compression to help it sound clean, intelligible, and fit into the overall mix. To get started, you can pretty much follow the steps I outlined in the first chapter of the book. EQ is almost always required and usually consists of a low-cut filter (set it around 80Hz), maybe some low-mid attenuation to reduce muddiness, and perhaps a bit of high-mid boost for additional presence and clarity. Lack of clarity and intelligibility is one of the more common problems I hear from an EQ perspective; give it a 2-3dB boost around 4kHz. On the other hand, some shows are so harsh and piercing it hurts to listen. This is too much signal energy around 3-5kHz (or higher), so try pulling that area down. Just don't overdo it or you'll make it sound like you threw a blanket over the host. On the bottom end, I will often add a small amount of low frequencies (roughly 240Hz, but it depends) to give a little beef and warmth to the voice.

Most DAWs provide built-in presets for EQ settings. These are good to get you started, especially if you haven't yet developed an ear for hearing EQ changes. Try the "male voice deep", "female voice bright", or whatever they have; maybe it'll work as-is or do fine with a little tweaking. I generally find presets to be somewhat overdone and end up reducing the gain amounts.

If your voice is sizzling the "s" sounds, then you'll need a de-esser. We're talking about *sibilance*, and it can get nasty and annoying. A de-esser is essentially a compressor tuned to a specific frequency region. You can probably just insert the plug-in and leave it be, although as your ear gets better you can hear how to adjust the target frequency and set how much reduction you want. Too much and you'll start sounding like a mush-mouth, which is probably not a good thing. The de-esser should go before any EQ.

Audio example 29: Sibilance

I hear lots of podcast interviews where the guy phoning it in sounds really muffled in comparison to the host sitting in her studio; it's hard to understand what that dude is saying while I'm driving down the freeway. Try to match tone quality as best you can to make it easier to hear both sides of the conversation. So, in this example the remote guest needs more clarity, which again is in the high-mid range on an EQ. Boost a little around 4k and it should sound brighter and more intelligible. You might also have to reduce the low-mids a little (somewhere in the 250-400 range) to reduce the muffled portion. You can't just leave it the way it is; it's too hard to listen and follow along.

Audio example 30: Muddy voice track vs low-mid attenuation for clarity

---

Finally, in many cases a touch of high frequency boost can give a voice more "air" and detail. We're talking roughly 8-10kHz and above, so set the EQ band to shelving mode, move the frequency select to 10k, then turn it up one or two dB. That's it, no more. A shelving EQ band will begin somewhere below the point where you set it, then provide constant gain (or attenuation if that's what you want) all the way up the rest of the frequency scale.

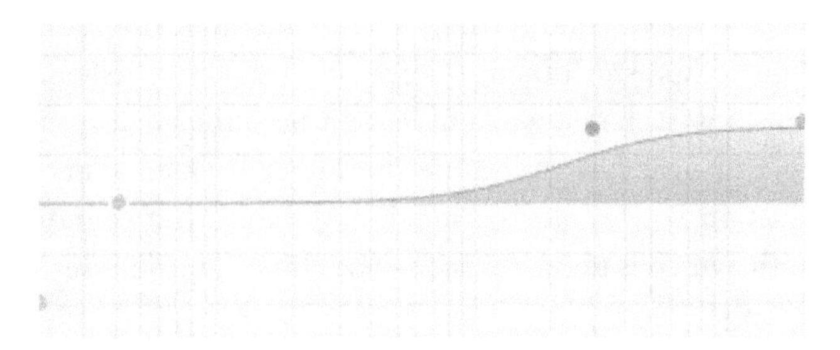

High frequency shelving boost

A few dB of compression will help even out the dynamics, meaning less variation in volume as they talk. Set the compressor's ratio to 3:1 and lower the threshold until it starts flashing 3-4dB of reduction. If you have more than one person on the show, put a compressor on each voice track; you'll adjust the threshold differently depending on their speech patterns and how animated they are. Many compressors have auto attack and release, so leave that turned on to make it easy on you. If not, set a not-quite-the-fastest attack time along with a medium release. The compressor chapter will explain further.

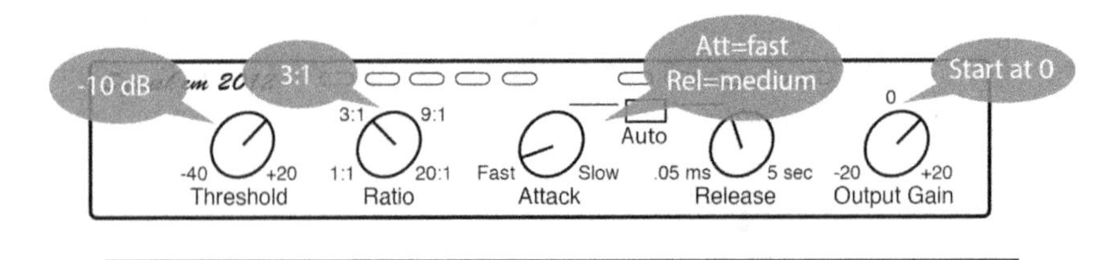

Audio example 31: Moderate vs too much compression

That's really all you need, but you can spice it up even more if you want to experiment. After I put a de-esser, EQ, and compressor on my tracks, I'll reach for a "shaping" EQ. Processors such as EQ and compressors come in many different flavors; some are more surgical and clean, others have more impact on the tone. I usually pick a really fat-sounding compressor that's modeled on one of the classics; it warms up the voice nicely. After fixing issues with an initial "clean-sounding" EQ, I'll then run it through another classic model that adds its own sheen and flavor, using very gentle settings such as a couple dB high and low frequency boost. If you really want to go wild with all this, start looking at stuff like saturators, harmonic enhancers, and so on. Don't go crazy with these, though, or it'll sound terrible.

Compressor on left and shaping EQ on right

EQ and compression take time to understand and learn how to use. It takes awhile just to start hearing what's going on, so review chapter one, read the following chapters that provide more detailed explanations, listen to the audio examples on the website, and keep practicing. Keep in mind, however, that poor mic selection and technique can only be improved to a certain extent. It's far, far better to learn how to record it correctly to begin with. Then all you have to do is sweeten it a bit and you're done.

Audio example 32: Original voice recording vs final processed track

By the way, unlike music production, you almost never want to add reverb to a podcast recording. Doing so muddies the sound, lessens intelligibility, and sounds unprofessional.

Stereo panning

Mono vs stereo? If your show is mostly voice with minimal music, then keep it mono. Your final show file will be much smaller (half), reducing your hosting costs over time. If music is important, and especially if you're running sound effects, creating a radio drama, or whatever, then go stereo. Assuming this is a stereo show, voices should always be panned in the center. It's annoying to listen in the car and hear the co-host coming all the way over from the passenger door speaker. Music should always be a stereo file; spread it out left and right, which leaves room in the middle for the host to be more easily heard. (For those of you paying attention, panning music this way doesn't actually open a silent hole in the middle, but the signal level is down slightly. We do this in music production to make "room" for lead vocals that are usually in the center.)

If the show has sound effects, think about how to use the stereo space creatively. My show has a jungle/tiki theme, and there are various environmental sounds that are spread all around.

Audio example 33: Stereo show music with jungle sounds

Volume balance

Volume balance is crucial, and apparently few podcasters ever think much about it. You don't want anyone sticking out or getting lost; an abruptly loud transition is jarring, and leaving somebody too quiet makes it too hard to hear what they're saying. We want all participants to sound equally loud and present in the show. The intro show music should also smoothly segue into the dialog; there's one show that shall go unnamed where the opening music is incredibly loud and distorted, then suddenly disappears to reveal these tiny voices. Yikes.

Audio example 34: Poor balance between show hosts

Audio example 35: Better balance

Audio example 36: Good balance

This can be as easy as adjusting all your faders at the beginning, assuming there are no big changes in volume throughout the show. Sometimes this needs more refinement, so we do this one segment at a time. Let's say you recorded the intro separately from the main conversation, and it's a tad louder for whatever reason. Each of these chunks is called a region or clip, and you can select a single clip and increase or attenuate its level. Pro Tools provides a tiny little fader for each clip where you just grab it and go up or down. Very quick and easy.

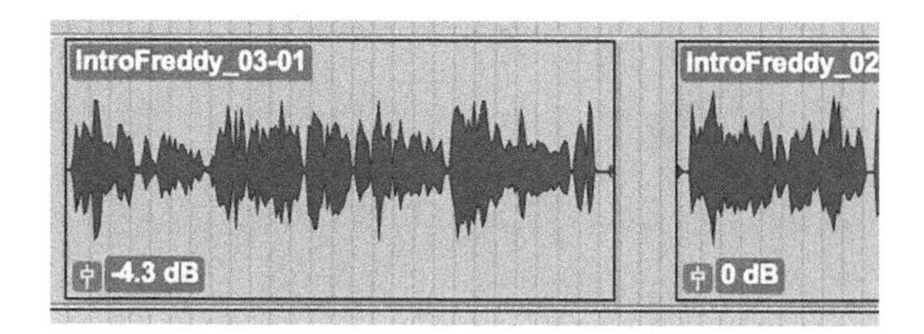

Clip gain to match relative volume

Let's say there's a spot where the host just loses it with a big huge shout; slice this region and reduce gain so it's more in line with the rest of the dialog. Make sure you apply a cross-fade at the edit points which will smooth the transition.

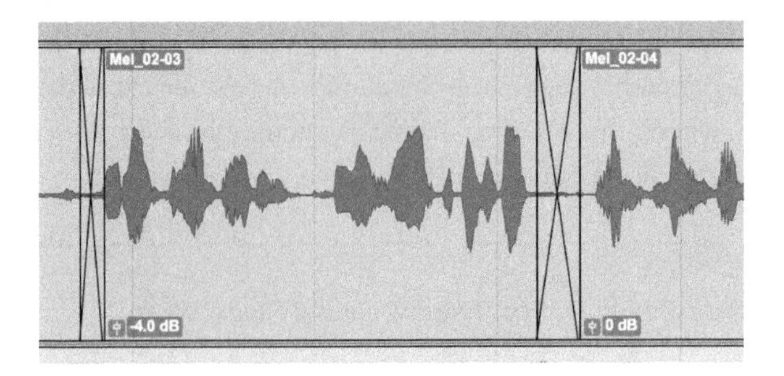

Select loud region, splice into separate clip, reduce gain, then crossfade to smooth the edit points

Sometimes more extensive work is required, depending on the nature of the show, how it was recorded, and how much effort you want to put into it. Let's say the show consists of an interview between two or three people who are in the same room. Each person has their own mic, and they're all sitting around the table. Go back and listen to one mic solo'd in the DAW, then turn the other mics on. You'll hear the voice change and sound more "roomy". Each mic is picking up everything in the room, including the other voices at the table. So when one person is talking, the other mics are capturing the sound, but from a greater distance and therefore with more room acoustics. If you decide this is too noticeable and needs to be fixed, or if you just want to clean up tracks when someone's not talking, there are a couple of approaches to try.

Advanced DAW applications like Pro Tools and Logic Pro have a feature called "strip silence". You set a threshold, which means anything quieter than this will be eliminated. What's left should be the actual talking from that person. The trick is that in a conversation where everyone is in the same room there are all kinds of noises made in response to each other, such as "mmm", "uh huh", and short bursts of chuckling or laughter. It's often very difficult to set a silence threshold that distinguishes these from leakage of the other voices in the room, so if strip silence is too aggressive you can try a noise gate.

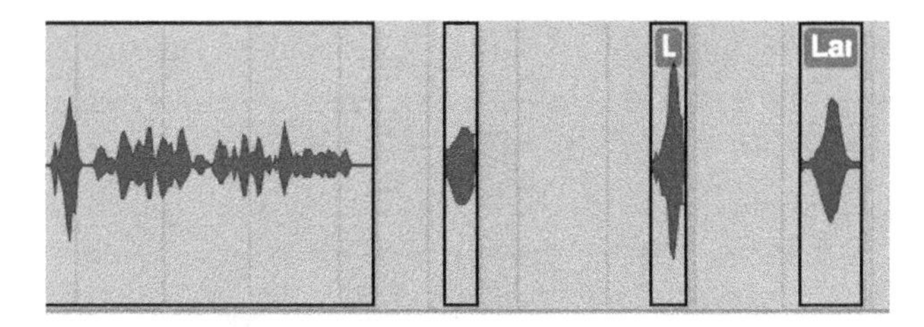

Strip silence leaves individual clips

Gates are processors applied to individual tracks that reduce volume automatically when a signal falls below a certain level. You're talking during the intro, ask your guest a question, then wait for the answer. A gate will open up the channel while you're talking, but as your last word fades away it'll kick in and attenuate the channel so you don't hear the background noise. Unlike strip silence, a gate is variable in how severely it shuts down this noise. It can be set to effectively turn off the channel, like strip silence, but it could also merely reduce the track volume a bit. This might be enough to lower distracting sounds while allowing interjections and such to come through. It's tricky, though, and I'm never quite happy with the results. Listen to episode seven of my show, The Themed Attraction Podcast (interview with Ryan Harmon); I used gates on the three participants. Maybe I'm too picky, but it's not smooth enough for me; most people won't notice or care. Give it a go and see what you think, but you'd better listen to your entire show to make sure it's not cutting out important stuff (you know, like actual words).

Yet another option is to automate the track volume throughout the show, turning the fader up and down as you go along. The idea is to set the volume level up when the person is talking, down when not, but allow for those low-level interactions that can be important. The most effective and accurate way to achieve this is by manual adjustments, where you write in the level changes rather than recording a fader move. The diagram below shows level automation between tracks. See the fluctuating line below the track's waveform? That's the volume, and sometimes we set it to pull the fader all the way down, other times just to reduce it a bit so it's not a complete mute. You'll also notice "fades", where the line doesn't just suddenly turn on, but ramps up and down. This smooths the transition, and the key is listening to all tracks together to see how each contributes. Recording automation like this is far more time consuming than using a gate or strip silence (or especially just leaving it as-is), so you have to decide what's worth your time for the desired show quality.

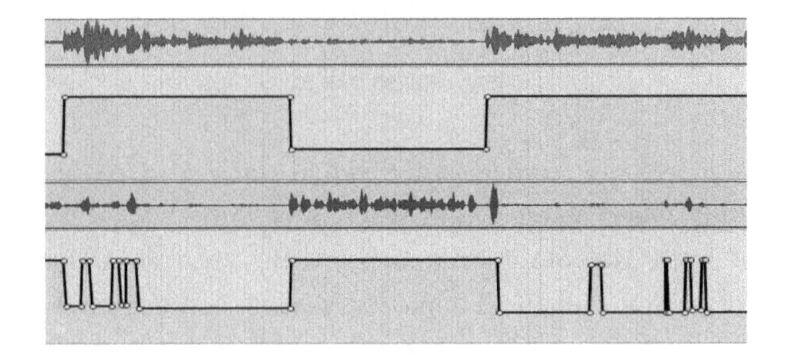

Audio example 37: Volume automation between show hosts (off, then on)

Now, if everyone is in different rooms or locations, this becomes much easier since there's no leakage between mics. Try the strip silence approach or just listen to everything together and see if you need to do anything at all (I don't ever use strip silence, but many podcasters do). Try all of these and see what works for you.

One more thing—if your session doesn't have one, create a stereo *master fader* (Logic Pro includes a *stereo output* by default in all sessions). This is the final "track" that everything feeds, so you use the meter to see how your overall show levels are running. This is crucial to ensure you're not going too high and clipping the final file (distortion). For those enterprising audio engineers out there, this is also where you could insert a bus compressor, something we use in music all the time. This is a special processor that "glues" the mix together; it sounds a bit tighter and punchier. Set it so it's barely registering maybe a dB or two of reduction. This isn't so important for podcasts, but it's a nice touch, especially if you have good show music. Don't worry if you skip this step—probably 98% of all other podcasters are right there with you.

All tracks feed to the master on the right

Once you've gone through all of these steps, go back and play the file at transition points—when one person takes over talking from another, going from music to dialog (and vice versa), your ad insert, and so on. You're listening for bad edits, rough transitions, volume imbalances, and anything else that will detract from your listeners' experience. Mix engineers in the music industry have an old trick where they turn their monitor volume down almost all the way; if anything sticks out then it's probably too loud in the mix. If you're satisfied, move on to the final stages: mastering and show file preparation.

Mastering

The mastering stage in music production involves taking a mixed song and tweaking it for incremental improvements. For a podcast, we want to make sure that the final show file plays back at the appropriate volume without clipping (distorting) and is easy to hear in noisy environments. This means regulating overall volume changes and setting it to the loudness standards that have been loosely adopted in the podcast community.

I'm going to step through three methods for this; the first two require plug-ins you may or may not already have. The third isn't nearly as good, but it's far better than nothing.

Method 1

You should have already ensured that no point in your show is clipping the master output. Insert a *peak limiter* on the master track; peak limiters take lower-level audio and increase it while maintaining a maximum ceiling on the overall track. By squashing dynamic range into a smaller space, the final show file sounds louder and fuller; overdo it and it starts sounding like hash. Like I said in the opening chapter, it's like a compressor on steroids, but it's not the same thing; you can't just crank up your standard audio compressor.

The sorry history behind this little gem comes from the music industry a couple decades ago, when record executives and artists decided that their album just *had* to be louder than everybody else's. The flawed reasoning is that a louder track must sound better. That might be true to a point, but they went crazy with it, forcing mastering engineers—at the threat of taking away their espresso machines—to crunch it. The loudness wars had begun, and it continued its madness until fans started buying records that sounded *terrible*. And they rebelled, gradually bringing us back to our senses. Go back and listen to a record from the 70s or 80s...not one that was re-mastered in the next twenty years, but an original. It sounds quieter than current stuff, but it has—get this—dynamics, space, clarity, musical-ness. Look at the waveform for that recording and it's got gobs of space between the various peaks and max zero, with lots of swings along the way. Compare that to an album that's over-compressed and looks like a solid mass that never lets up. It's not pleasant listening to something like that, so we want to balance ensuring our recordings sound relatively loud, but not crunched.

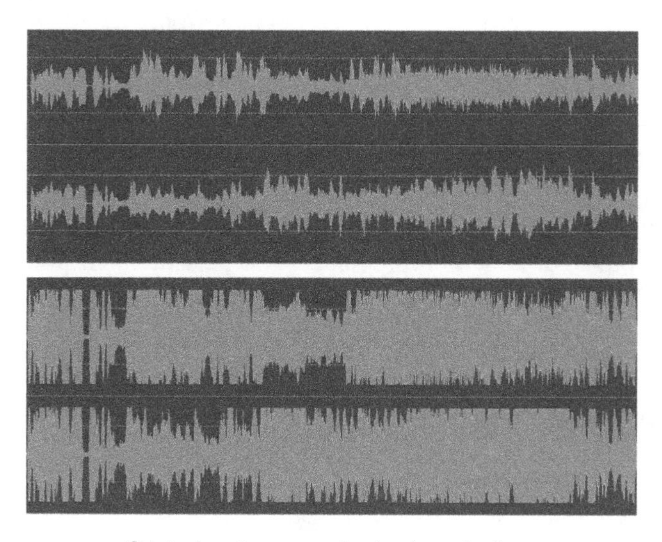

Original on top, squashed pulp on bottom

Anyway, back to our story. Be moderate with the settings. Pro Tools comes with a peak limiter called *Maxim* (*adaptive limiter* in Logic Pro); set the output peak level for -1.5 dB and slowly decrease the threshold while playing the track. Listen as it gets louder, then progressively more nasty; find a balance that sounds louder but still natural. Many podcasters overdo this and their shows sound loud, but distorted and edgy. The -1.5 output ceiling is there to protect the file from clipping; anything that flashes beyond zero is instantly distorted in a digital system. We maintain this buffer of 1 or 2dB so the algorithms used for encoding mp3s have some headroom in which to work.

Audio example 38: Peak limiter bypassed

Audio example 39: Peak limiter moderate setting

Audio example 40: Peak limiter overdone

Method 2

My preferred way, if you can get a loudness control processor such as RX Loudness Control from iZotope or WLM Plus from Waves, is to avoid the peak limiter altogether. There's a recent trend in the audio world for measuring deliverable files in terms of *loudness*, not absolute signal level. Loudness is a subjective concept where we tend to average dynamic changes over time. The goal is to deliver audio that sounds about the same loudness as other files, whether that's other podcast shows, albums, or television programs. The measurement units for this are LUFS (also known as LKFS). The "LU" stands for *loudness unit*, referenced to full scale. There are adopted loudness standards for broadcast radio, Netflix, music streaming, etc. The podcast community has generally accepted -18LUFS (mono) and -16LUFS (stereo).

We can largely thank the music streaming services for bringing sanity to the loudness wars. No matter what you send them in terms of a master file, their system will automatically adjust the relative loudness to their standards, thus ensuring that people's listening experiences are consistent and fluid from one album to the next. So there's no need to crank your master as far as it will go; they'll just turn it back down. Produce a great sounding track with nice dynamics and it will translate correctly.

The way I do this (for now...workflows evolve) is to bounce my mix from Pro Tools, then import this stereo master into a template session set up with my loudness tools all set. RX Loudness Control from iZotope is an audiosuite plug-in, which works a bit differently than inserting a compressor or EQ on a track. Highlight the entire waveform, then call up the processor. Set your target to -16LUFS, true peak to -2dBTP, make sure the gate is enabled, and click *analyze*. It will scan the entire file and return with the results: overall loudness, short term loudness, loudness range, and maximum peak level. Now click *render* and it'll do its magic, adjusting things so as to make the entire file conform to your desired settings.

Some podcasters will use loudness meters on individual tracks, primarily to see how those levels are running and perhaps adjust individual clips along the way to get them close. You'll have to experiment over time and get a feel for what works for you. Regardless, a loudness plug-in is required on the overall mix to set the final level of the entire show.

In the screenshot below, the loudness plug-in has measured the show's long-term level at -24.3LUFS with an average dynamic range of 9.8. Larger dynamic range is good for music, smaller is generally better for dialog. The lower half of the plug-in is where you set the target ranges for processing, in this case the standard -16LUFS for stereo podcasts with a -2dB true peak maximum to prevent distortion.

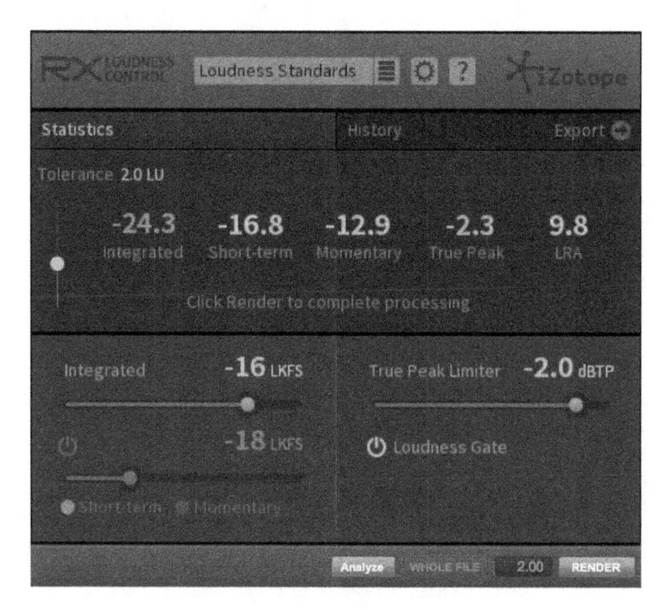

Loudness target set at -16LUFS and -2dB max peak

If you've done a good job mixing the original show session, meaning balancing your different tracks, judicious use of compression, and using volume automation to reduce hot spots and bring up low levels, then the final file will sound natural and consistent. And that's the trick. Spend time fine-tuning your show from the beginning and it'll pay off here.

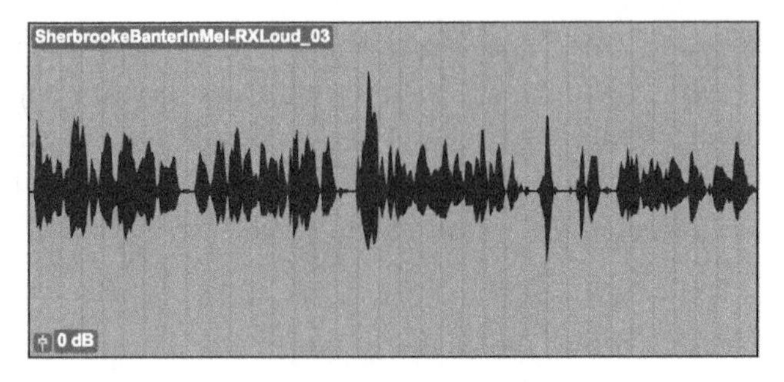

Original waveform

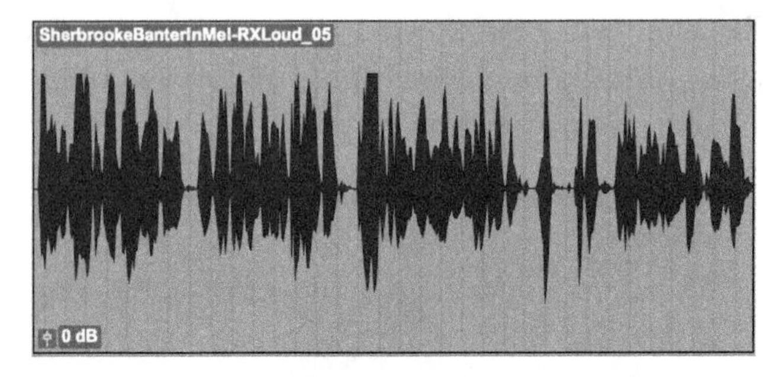

Loudness processing

Method 3

If you don't have a peak limiter or loudness plug-in, at the very least *normalize* the final WAV file. This process examines the entire audio track, searching for the highest peak level. Let's say your show hit -5dB at the highest point. This value is subtracted from 0dB, which is absolute max, giving us a difference of 4dB. We want to preserve a buffer below zero, giving the mp3 algorithm some space to work its magic, so set the max ceiling to -1.5dB (some people use -1, others -2, it varies). This process will then increase the entire show file 2.5dB, making it a little louder across the board. It won't change any dynamics or relative loudness throughout the show; it only increases (or decreases, if the original mix was too high) the existing audio as-is. A dB or two makes a tiny difference, but I hear some shows that are so low I have to crank the volume nearly twice as high.

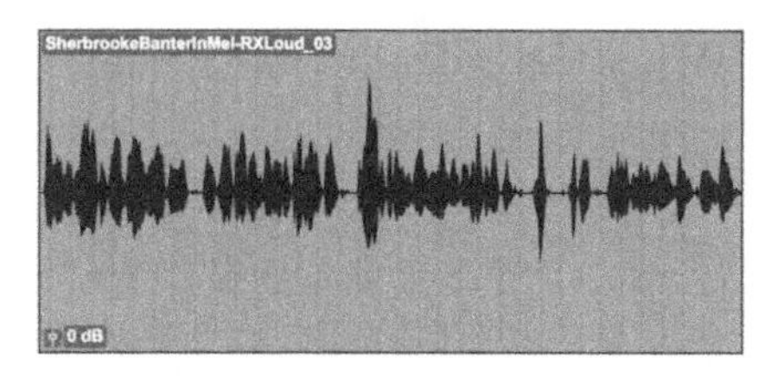

Original waveform

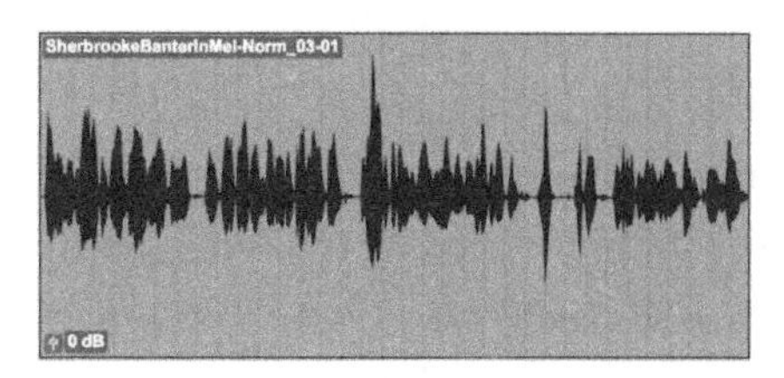

Normalized

The final master

When everything is ready, the show has to be exported as a final audio file, ready for uploading to a podcast hosting service. Even though you always want to record and edit uncompressed, high quality WAV files, the finished product must be compressed and in a format that works on listeners' media players. Most podcasts are distributed as an mp3, a highly-compressed format that reduces large CD-quality files to a fraction of their size. They work quite well for predominantly voice productions like this. The other option to consider is the slightly better AAC (m4a), which was adopted by Apple for their iTunes/Apple Music service. Most players can handle either, though mp3 is guaranteed to work pretty much everywhere.

Use the following settings when exporting (bouncing) this final file:

- WAV
- Same sample rate as your original session (44.1kHz is what we specified earlier)
- 24-bit
- Include mp3

You want both WAV and mp3 files; the WAV is for archiving, the mp3 is for uploading to your host provider. If your DAW won't do both, export the WAV first, then use that to create an mp3. Set the mp3 encoding rate to 96kbit/s. You can get away with 64 for mono, and I would suggest 128 or higher for stereo (podcasting, not music—you need 256 for decent sounding stereo music files). The higher the rate, the larger the file, which translates directly into higher costs for you as many hosting services charge by data uploads. For our show we commissioned custom, high quality music; I want it to sound as good as possible without unduly forcing listeners to stream overly large files, so I run 160. An hour-long stereo show encoded at 160kbit/s is roughly 72M; a mono show would be half that. Experiment with different settings, load it into your phone, and go jogging to see how each compares. Or in my case go sit on the couch and cycle through them with a good cup of coffee.

A dialog box will appear asking for various mp3 metadata tagging information; go ahead and fill that out as it helps the end listener know more about the show, find it in their catalog, and so on. You probably want to enter the title of each episode in the title field, along with the show number and/or date, and use the album slot for the main podcast show title. This sorts everything together in podcast players.

A better mp3 option

A more powerful way to generate a final mp3 show file is to use a software application such as Forecast. This (free) program puts everything together for you. Start by importing your WAV master file and it'll start crunching an mp3; meantime add show notes and title information, chapter markers, and your show icon, and it'll generate all the data required by iTunes for cataloging in its database. Save your file and it's ready for uploading to your hosting service.

Before you move on...

Have you listened to your final show file? Away from the studio (or bedroom)? It's obvious many podcasters have not, or at least not with a critical ear. There's no excuse for publicly distributing a file that has easily fixable issues. I always take my mp3 and play it in the car on the way home (it's a long commute) to make sure I haven't missed something, which can happen for a long show. I guess this depends on the nature of your particular show and the quality that seems important, but yikes, this is the same as proof-reading your marketing materials or resume. You gotta do it, at least for awhile to make sure things are in the ballpark.

Posting and hosting

For people to find, subscribe, and download your shows, there's a particular way they have to be made available. Most podcast players look to the iTunes/Apple Podcasts database for a listing of shows and episodes, so you've got to be registered there. But the show file itself has to be hosted on somebody's server. The easiest route is to buy a monthly subscription with a hosting service such as Blubrry or Libsyn; different pricing tiers provide increasing levels of data uploads and other variables. Search around to see who the currently popular services are and follow their instructions. But to test or enjoy your own show, remember it's just an audio file. Copy it to your music library and play away.

FOUR
SHOW DESIGN AND QUALITY ISSUES

I GET IT—MOST podcasters are not professionals. But the better your show sounds, the more seriously people will view it, respect what it's about, and help spread the word. So don't take any of this personally; use it to help you get better. Most of this doesn't take much time or cost any money once you're aware of it.

When's the show ever going to start?

The introduction to your show is meant to grab attention and set things up for you to make your appearance in glorious fashion. It doesn't take ninety seconds or longer to do that; too many intros run way, way too long. Just get to it, please. After they've listened to a handful of episodes that cool song or whatever can get really old if it doesn't duck out at a reasonable point.

Can't hear the show without cranking it up super high

Do you need to crank the volume twice as much to hear your show? Then go back to the mixing chapter and review the three options I provided: peak limiter, loudness control, or normalizing. Any of these will produce a file that's relatively loud. If your original recording levels were low and you didn't do anything to adjust, then the final show file will be too quiet.

I can't hear that other dude

I just listened to an episode this morning where I couldn't for the life of me hear the co-host. He was way quieter than the main host. No reason for this—in this case each person was recorded separately in their own locations. Move those track faders (yes, there's a reason they're made that way) so they sound about the same volume.

My guest keeps trailing off as she speaks where it's tough to hear everything

People's speech can vary quite a bit between loud, confident phrases and quieter moments when they're more reflective, thinking about what they want to say. It's helpful to even this out somewhat, and a compressor can improve this to a point. You don't want to smash the loud sections too much, though, so find a moderate balance. This can also be automated, and sometimes I'll draw in a slight automation volume curve to raise the end of a particularly difficult-to-hear sentence. If you can splurge for the iZotope RX package, the advanced version has a dialog leveler that does wonders for this sort of issue; it evens out speech very smoothly.

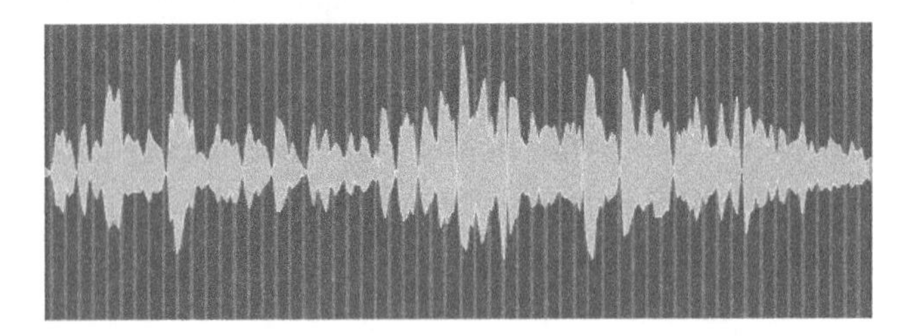

Before dialog leveling; see variation in waveform

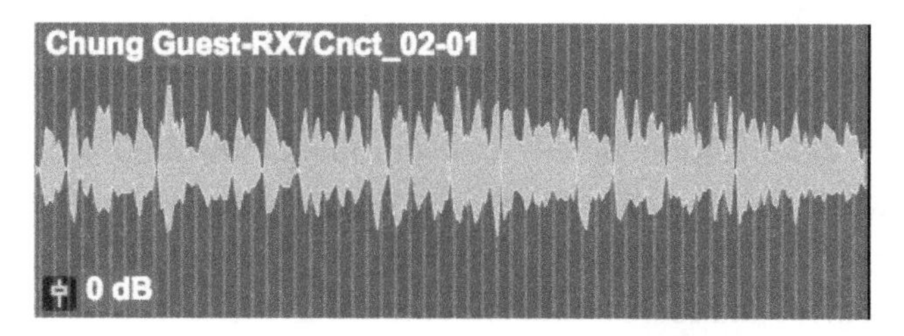

After dialog leveling; more consistent waveform

Whoa, that was abrupt

Carefully listen throughout the show to ensure everything transitions smoothly. Abrupt volume differences, cut-offs, and so on are extremely annoying. Again, work those faders and also apply short fade ins and outs between adjoining segments.

That host is awfully distorted and noisy

Back to the basics: a decent mic placed in the right spot with appropriate recording levels will fix this. At all costs make sure that needle is not coming close to max zero on the meter. And use a decent mic. Always.

I forgot to check recording and bounce levels and my final show is really distorted

Yeah, well, too bad. Not much you can do about this, although again, iZotope can come to the rescue with their de-clipping module. It's pretty good, but the free, easy fix is to be really careful from the beginning. Lesson learned.

Sounds like a laptop or cheap headset mic

Basics again. Don't use those mics. Ever. Anyone willing to pay the monthly hosting fees for their show can afford a decent USB mic.

The host sounds like they're in a cave

Did we mention basics? Sit close to your mic as we discussed earlier. If you're two or three feet away (or more) it's not gonna sound good. Especially if you're in the kitchen or a bedroom with lots of tile or bare walls. Mic placement and room treatment doesn't have to cost anything.

Everybody in the room can be heard in each mic

If there's more than one person talking in the room during a podcast session, the mics will inevitably pick up the *other* person in addition to the one it's intended for. This is called leakage, and while you can't eliminate it entirely in this situation, you can mitigate its effect. Avoid having people sit close to each other—spread them around the table facing one another. Then the microphones are facing away from the other sources, reducing how much leakage they pick up. Use dynamic mics instead of condensers; dynamics are far less sensitive to sound beyond the immediate capsule area. Bare, untreated walls will increase leakage since everyone's voices are reflecting around the room; install acoustic panels, hang thick curtains, do anything to break up the plain, flat surfaces. Finally, ensure that each person is fairly close to their mic; greater distances require the mic preamp to be increased more, which brings in more of the surrounding room environment. The professional shows sound so good partly because each person is comfortably tucked away in their very own, acoustically-designed studio. Can't have leakage problems if no one else is in the room.

My voice sounds muddy or boomy

With certain types of microphones, close placement to the sound source results in a low-frequency boost called *proximity effect*. This occurs with directional mics such as cardioid (uni-directional), hyper-cardioid, super-cardioid, and bi-directional. Omni-directional mics are not affected by this. Simply move back a tad. You can also insert a low-cut filter and set it to around 80Hz or so.

Muddiness is slightly higher in the frequency range, roughly 250-

400Hz depending on the source. Experiment with an EQ to find a particularly cloudy region and decrease it. Turn up the gain for the low-mid band at least 6 dB or so. Now rotate the frequency select control and listen for the changing sound of the frequencies as you move up and down the scale. Turn back and forth until you hear a region which sounds less distinct and muddy compared to the rest. Now reduce the boost control back to a negative number, perhaps -3dB or more. The more you cut, the more of the overall sound will be taken out, possibly thinning the sound too much. In this case use the bandwidth control (Q) to narrow the frequency region being effected.

What's that sound in the background?

Any background noise that's not part of the show should be reduced as much as possible, preferably during tracking. Air conditioning vents are a common problem, along with other building noises, traffic outside, ringing phones, and so on. Face mics away from vents as much as possible, or even consider turning the AC off during each take. Make sure everyone turns their phone off and removes clanking jewelry and anything else that'll rattle and be annoying.

Too much reverb

Most people do not have a decent room in which to record. And they keep forgetting some of the simple, free tips I provided earlier in the book. If your recorded track has significant amounts of reverb on it (echoey sounding), then the only thing you can do is find a plug-in that is designed to reduce reverb. iZotope, of course, has one, so put the plug-in first thing on the track (before all other processors) and experiment with the settings. Too much processing generates digital artifacts, which sound like aliens taking over the show (electronic-sounding weirdness). It's a balance. Of course there are times when you can't help where you're recording, but if this is your regular space, fix it. It's worth the effort.

The guest called and wants three sentences removed from the middle!

Often the need arises to remove a portion of a conversation; maybe someone said something regretful, changed their mind, whatever. This can usually be done, but making it sound seamless takes a bit of practice and patience. If there's space before the phrase begins, like a short pause between sentences, then it's easier. But often people finish a thought and immediately take a breath to launch into the next bit of inspired podcasting goodness. This can be very tricky or perhaps impossible. Experiment with various break points, remove the unwanted portion, and slide the remaining region over to close the hole. If there are multiple tracks this has to be done all together so as to keep everyone in sync. Then adjust the boundary as needed, applying a cross-fade to smooth over the joint. Here's a fairly complex situation where it took some fiddling to get the first edit point smooth and transparent.

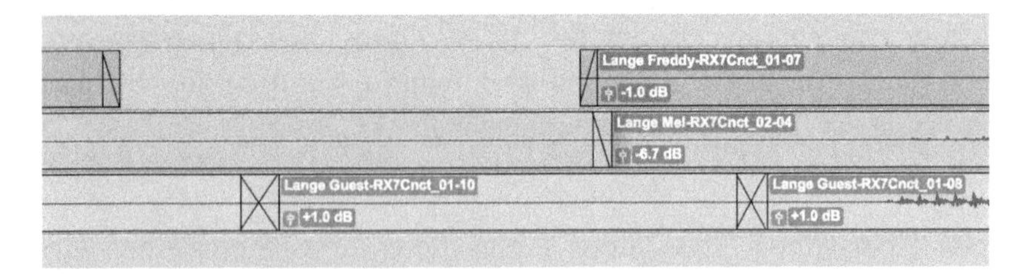

Staggered edit points to cover the breaks

My guest rocked back and forth in her chair the whole time and everything else bled into her mic

It's okay to give instructions to your guests; we can safely assume the majority of them are not professionals at this sort of thing. In this case, sitting too far back meant the mic had to be turned up more and was picking up the entire room. Close to the mic means it'll mainly capture the voice. Tell your guests how to sit close to the mic and not move around, to not touch the mic or stand, cough away from the mic, and so on. Some podcasters will use good quality headset mics, like those you see on people giving TED talks and such. I'm not sure I'd use it for my primary hosting dialog, but for an interview they can work quite well. But connect them with a cable; don't go wireless.

Music should be intentional

Nearly everybody uses music in their shows, and for good reason. But the choice of music, how long it lasts, and where it goes should be very carefully considered. Too often I hear random stuff that has little point except to check the box that "we want music in our show". Either it doesn't match the tone or subject of the show and doesn't contribute to the story, or it pops up here and there for little reason, or even runs in the background the entire length of the show. Make your music choices intentional—use sparingly, pick that special clip that works in that particular spot, and watch the volume on how it comes in and goes out.

Music should be legal

No, you can't just find your favorite Disney soundtrack and play it as your theme song. Search for royalty-free music online or buy a track from a place that sells music and sounds in exchange for the right to use them in your productions. It doesn't matter whether you're making money from the show or not; this is a public distribution of a copyrighted work which you don't own. For my show we hired a friend who composes in Hollywood for a living; he got a little cash and on-going credits on the show and we scored (pun intended) a first-rate, awesome music track. Our closing music is from another friend's album; they were agreeable to letting us use it in exchange for giving them a shout-out and credits. Just do it legally, please. Then you don't have to go through life looking over your shoulder...

Learn from the professionals

Want to get better at this? Want your show to sound like something more than a bedroom hobby? Do what any aspiring filmmaker, or writer, or photographer, or any other creative would do when learning a new craft—observe the masters and break down how they do it. Listen to the top of the line podcasts, such as from NPR and affiliates (lots more to choose from, of course), and model your work after them. This doesn't mean copy their ideas to the letter, but take notes on their show design, format, use of music, sound effects, transitions, ad placement, introductions, and so on. It's a show, not a meandering stream of thought that you simply throw up on the site. This podcast thing has rapidly become a serious enterprise, and the days where you can be a hot commodity running a terrible show are fading fast.

FIVE
EQUIPMENT OPTIONS

THIS CHAPTER IS tough because no matter what I include, some of it will be obsolete next week. Oh well, here goes.

Microphones

We've talked about the main things to consider when buying a mic. Dynamic or condenser? USB or XLR? Should I get a second mortgage on the house? The answer is it depends (like most everything in life). Yes, there are cheap mics that you should avoid, but there are exceptions to the price guidelines. And yes a vintage condenser will almost certainly sound better than that cheap USB, but there are so many variables it's a complicated thing. Or not. Just search around for reviews online and you'll get a sense of which mics podcasters are typically using. One of the best reviews I've found of mics and interfaces is from Marco Arment, software developer and active podcaster in the tech community. Start there at https://marco.org/podcasting-microphones#rank.

Of course he makes it clear these tests were based on *his* voice. It's different for different people. Most would enthusiastically agree that the Shure SM7B is one of the best podcasting mics you can get. It's not outrageously expensive ($350 or so), has a built-in pop filter, and looks really cool in front of your face. But he didn't like it. Neither did I. In my case I think it's because I don't have a deep voice, which would pair better with the SM7. Mics are a personal thing, so the challenge is that you really need to try different models. Of course you can't buy a handful and take 'em home, but see if your local audio dealer will let you try them out, even if it's at the store. Make some recordings and take them home to listen. Marco reinforces many of the differences we've already discussed between dynamics and condensers: dynamics are less sensitive, so they don't pick up the folks around you as much. You also get far less room sound (a good thing). But you need to stay fairly close and consistent to the mic. Condensers have wonderful articulation and detail, but they grab everything around them. Not very usable for multiple people in the room or if your recording space isn't exactly designed to be a recording space.

A few mics that have been popular over the years: Sennheiser's e835 dynamic is pretty common and inexpensive at $100. The Electro-Voice RE20 has been around for decades and is very popular in radio. It's a large diaphragm dynamic, like the SM7B, and has that deep DJ kind of sound. It's got a younger brother, the RE320, which is cheaper and sounds almost as good. I like the Audio Technica large diaphragm condensers, such as the AT4040, as they offer a smooth, rich sound; the $170 AT2020 USB is another good contender balancing the sound of a condenser with price. Of course any Neumann large diaphragm condenser is gold (and priced accordingly). I tried the Telefunken M82 on both of our show's co-hosts and it sounded terrible. But for some people it's been great. It's a large diaphragm dynamic which would also work well on kick drum (its original purpose). Others from Heil, Shure, Blue, and Rode offer a variety of sound quality and price points. AKG just released the Lyra, a $150 USB-C mic that's got a headphone jack and volume control, mute switch, adjustable polar pattern (from cardioid to a wide stereo spread for group settings), and looks so cool. In another year or so we'll have a whole lineup of microphones designed specifically for podcasters, rather than having to use a kick drum mic for our show. For now, start searching around and ask people you know who do voice recording.

Audio interface/mixer

An often overlooked consideration is that some mics require a hefty preamplifier to drive them. The output signal of any microphone is extremely low, say 40-60dB below what we call *line level*. Audio gear such as mixers, outboard processors, recorders, and so on run at line level, so we have to run a mic output through a preamplifier to crank it up. Most mics used for podcasting work fine with your average audio interface or mixer, but a few, such as the notorious SM7B, require a lot of drive from that preamp. Either you need to buy a serious interface, such as the Universal Audio Apollo Twin, or get a Cloudlifter. These little doodads go in-between the mic and the interface and increase the mic level up to 25dB. You need one for each mic being used, so at $150 a pop this could become prohibitive. In that case either pony up for a high end interface or get a different type of microphone.

Less expensive audio interfaces that have good reputations include models from Focusrite (recommended) and Presonus. A 2-channel Presonus starts around $100 with the entry level Focusrite just a bit higher at $110. Either of these will work fine. Some models have iOS connections for recording with iPads and iPhones. Of course there are other brands such as MOTU, Tascam, Line 6, and Zoom, and so on. You need one mic preamp for every microphone you're planning to record at one time. So if your shows include a host and three others around the table, a four mic input interface is required.

Another option is to buy a combination mixer/recorder. These offer multiple mic preamps and serve as the audio interface to the computer. Get one of the newer generation of podcast-specific mixer/recorders such as Zoom's Lifetrak L-8 and the RodeCaster Pro from Rode. These offer several mic preamps, built-in SD card/flash recording, USB to the computer, and even a TRRS phone jack for connecting a remote caller. (TRRS stands for tip-ring-ring-sleeve. The plug has three black rings around the shaft which separate audio left, audio right, ground, and microphone.) Trigger pads can be programmed with any audio sound you want: push a pad to start the intro music, tap another to signal a gong when your guest guesses wrong on the game show. Multiple headphone outputs with individual volume control is extremely useful when having a group discussion in the same room. The entire surface is laid out a bit differently than a standard mixer, making it easier and more efficient for podcasters to use. We'll see more of this sort of thing, where manufacturers continue to recognize the unique needs and marketability (read that "sales") of the podcast industry.

Recorders

We just mentioned the combination mixer/recorders that provide built-in recording capability. You can use them as an audio interface and record to your computer, but if that's not convenient simply transfer your SD card after you get back from that remote gig in the Everglades. Make sure the unit can record independent tracks, not just an overall stereo file; you want to process each voice separately during the mixing phase.

Handheld recorders are super convenient—they have built-in mics, SD card recording, and are portable. Many of these offer a couple external mic inputs; the Zoom H6 has the capability for up to six mics. If you're using the built-in mics, placement is an issue, so go with external mics that you can situate properly. When you go shopping be careful to note that some handhelds are intended for voice recording only, so the quality isn't as good. This might seem just fine with you, however, since you're recording, you know, voices, but these units are really intended for quick dictation purposes. The slight downside for portables is that it can be harder to keep an eye on recording levels as you can't see the meters as easily as on a computer screen. And if you're using the built-in mics, you can't adjust anything on the recorder, such as mic levels, without introducing noise into the recording when you touch it.

All of these devices, including regular audio interfaces, will connect to a computer for recording. The mixers and portable units, however, are more expensive than the lower end interfaces such as those from Focusrite and Presonus. If you're only going to be recording from your basement, go with the interface. If the plan is to record shows out in the field as well as laying down monologues at home, then the other units will serve both functions.

Many podcasters are recording straight to their iPads, even editing and mastering their shows in some cases. Ferrite Recording Studio has become very popular for this as it supports multitrack recording and provides solid tools for editing. With the iPad Pro we're finally getting better capability for accessing USB storage and accessories such as the Apple Pencil. If you're doing in-depth processing, though, you're better off transferring these files to your computer and putting the show together there. And remember you'll need enough mic preamplifiers on an iOS-compatible interface to match the number of mics you're recording at one time.

Mic stands

Many so-called podcasting microphones come with teeny, cheapo stands that do nothing more than hold the mic upright. That's too far away, so invest in a decent stand that will position the mic close to your face while sitting comfortably. You're gonna be there for hours, so get this right. Buy a quality ergonomic chair, set your gear and screens so you don't have to lean or strain your eyes, and then figure out what's needed to get the mic situated so you're not messing all that up. Usually this means an adjustable boom arm, which clamps to your desk and swings out wherever you want it. I've got the Telefunken M786 Radio Boom Arm and so far it's been great. You can attach any mic to it, pull it over when recording, then swing it out of the way when you're done. If $100 is too much, regular boom mic stands can be had for $20-30 each. This will work fine; it's just more cumbersome and in the way.

Several microphones ship with shock mounts, which is an elastic webbing that suspends the mic away from the physical mic stand. If yours has one, use it, so when you bump the table it won't thump as much into the recording.

Pop filters

Hold your hand in front of your face and talk—you'll feel puffs of air, especially with hard consonants like "p" and "b". That's what causes pops and booms in a mic, so do two things to reduce this. Position the mic slightly off to one side and perhaps a bit higher than your mouth. Certainly don't talk straight down into it; those air blasts are aimed downward. The next thing is to mount a pop filter in front of the mic without touching the grill. These are usually nylon or metal and are very effective at diffusing pop-inducing currents. I prefer the metal screens because they look more professional and will presumably last longer than nylon, but don't get hung up over this. There's nothing magic about it—just pick one that's not so expensive.

Headphones

I mentioned earlier that selecting headphones can be as challenging as finding the right mic. They all sound different, so you'll have to spend time getting used to whichever one you have. This means listening to lots of quality music and podcast shows to get a baseline to shoot for. Some phones are very bassy (I'm thinkin' of you, Beats), others overly bright, some with good clarity and definition, others less so. Once you know how your phones sound with professional recordings, that's your target for making your own show files. I've been using the Sony MDR-7506 for decades now. They're a touch bright, but I'm used to that. I've also tried the Audio Technica ATH-M50x, which sound great, and the Beyerdynamic DT 1990 Pro, which sound even better (they certainly ought to for the price tag). And yet I continue to pull on my Sonys every time I sit down to work on a show. So does the rest of the world—this is the model you'll see most often in professional work.

Don't use earbuds, especially the cheap ones that came with your iPhone. You need over-the-ear headphones with a balanced frequency response. Noise cancellation might be great for the flight home, but unnecessary for podcast work. Other good brands include Sennheiser and AKG; most of these companies make quality stuff along with cheaper models that aren't nearly as good. So price does make a difference, but don't go overboard, at least at first. The Sony 7506 runs about $90.

Wireless

Try to avoid wireless connections as much as possible; there are too many things that can go wrong. Wires almost always work (keep spares on hand). Wireless microphones are susceptible to dropouts, interference, battery issues (and cost), and don't sound quite as good as a cabled mic. Sure, there are situations where running cables to everyone is not advisable, so do what you need, but if you're buying the gear it can really add up. Quality wireless systems are several hundred dollars each.

By the way, this applies to bluetooth connections as well; while they're pretty reliable, why take a chance? Plug it in.

Software

Mixing and producing a final show file requires a DAW (digital audio workstation). This means the software has to record and mix multiple, independent tracks or parts while providing signal processing along the way. The heavies in this market are Pro Tools and Logic Pro, professional-level applications that'll do everything as well as cook breakfast for you. There are others that do the same work but are cheaper and arguably easier to use. Presonus' Studio One series is a popular choice these days for audio and music production. Owners of Adobe's Creative Suite already have the extremely capable Audition; Cubase is another option. GarageBand comes bundled with Apple computers and works fine for starters, but ought to be replaced as you get better at your craft for a more flexible, powerful tool. Audacity is everywhere primarily because it's free. But free comes at a cost—it's not elegant and doesn't quite work like a normal DAW. It'll handle basic recording and processing, however, so if that's what you can get your hands on go for it.

Pro Tools comes as a subscription model, requiring an on-going annual fee. Logic Pro is probably a better choice if you're not working in the pro audio industry; pay $199 one time and you're all set. A few audio interfaces come bundled with a DAW, such as those from Presonus, which ship with Studio One Artist.

All DAWs include a variety of signal processing (plug-ins), such as EQs, compressors, and effects. The more sophisticated processors such as noise reduction and loudness control often require separate add-on purchases. The iZotope RX Loudness Control that I described in the mastering chapter runs $349. The WLM Plus Loudness Meter from Waves will do almost the same thing and costs even more, but they regularly run sales and so I got my copy for $70. The main difference is that RX Loudness will automatically adjust the audio file so as to hit the target range, whereas WLM requires a manual adjustment while watching the meters. Not a big deal as long as you can get it for the cheaper price. For noise reduction and other sophisticated problem-solving tools, the shoot-for-the-moon iZotope RX Advanced package will set you back $1200 (the basic Essentials version includes voice de-noise, however, so for $129 that might be a viable starter option). See what comes with your DAW; it might already have some of these things. Adobe Audition, for example, ships with noise reduction and loudness tools. Other functions I've found to be helpful for podcast processing include de-essers (to reduce excessive sibilance), de-reverb (reduce surrounding reverberation), and voice de-noise (reduce other noises besides the actual voice recording). These typically require some experimentation to find a balance; the goal is to improve the track, but too much processing starts to sound really weird.

All DAWs can export an mp3 file, but my favorite podcast show file application is Forecast. This free tool will import the mixed WAV file and create the mp3 while you enter all the necessary show information for the Apple Podcasts directory, including the show graphic logo. It'll even encode chapters if you want to set this for your episodes.

So what exactly do you need to get at the store on the way home? Here's the basic list:

- Microphone(s)
- Pop filter
- Mic stand
- Mic cable (with spares)
- Headphones
- Audio interface, portable recorder, or mixer/recorder
- Computer/iPad
- Software
- Snacks

There are lots of options these days for equipment. Some of it is inexpensive but pretty decent, some of it is awesome and practically unaffordable for mere mortals, and much of it is, well, cheap. My motto is to always buy as high quality as I can afford. It'll last longer, work better in the meantime, and just feels good to use. Splurge on a decent mic and interface, then upgrade your software later. When your show starts generating tons of Patreon dollars you can step up to that classic Neumann...or just buy a Porsche instead.

SIX

MICROPHONES

THE NEXT FEW chapters are for those of you who can't get enough of this stuff and want to know more about how audio works and/or can't get to sleep at night. This'll fix both conditions.

All microphones are *transducers*, which simply means they are devices that transfer energy from one system to another. Microphones take acoustic sound waves and generate electrical signals. Other transducers in audio include tape recorders, speakers, guitar pickups, and phono cartridges for record players.

Although the sole purpose of a microphone is to capture an acoustic event and convert it to an electrical signal, there are multiple ways to achieve this. None is particularly any better than another, and variation in microphone design is deliberate so as to provide each model with a distinct sound. The result is a wide variety of mics to choose from so you can find the perfect match for your golden voice.

The two primary design types are *dynamic* and *condenser*. Other variables such as component materials and circuit design will also contribute to each mic's distinctive sound.

Dynamic microphones

This design employs the physics principle of electromagnetic induction. Inside the capsule of the mic a coil of wire is suspended within a magnetic field. The microphone diaphragm is attached to the coil, and when sound waves hit and move the diaphragm, the coil of wire also moves back and forth within the magnetic field. This generates an output voltage that varies according to the diaphragm movement. So, the output signal varies analogous to the changes in the original acoustic sound wave...and yes, that's where we get the term *analog*.

Examples of dynamic microphones include the Sennheiser e835, Shure SM7B, Electro-Voice RE20, and Heil PR40.

There are two variations of dynamic microphones:

- *Moving coil*: This is the most common type of dynamic microphone.
- *Ribbon*: Instead of a coil of wire, a thin ribbon of electrically conductive material is suspended within a magnetic field. The principle is the same as for moving coil, although since sound can strike the ribbon on both sides these mics are inherently bi-directional. These are very popular for studio recording due to their warm, smooth response, but not practical for podcasting.

Condenser microphones

The principle of electrostatics is the basis for condenser mics, which are also known as capacitor microphones. These have two electrically charged parallel plates to transduce acoustic waves; one is movable (the diaphragm) and the other is fixed, effectively forming a capacitor. When sound waves hit the movable plate, the distance between the two plates varies. This changes the capacitance, generating a corresponding output current.

Condenser mics require a DC power source to charge the plates and power an internal preamp; this low-level amplifier reduces high internal impedance and increases the output level a bit. This is not the same as the mic preamp on your interface that boosts the mic signal up to standard line level. The power source is called phantom power (+48V) and is supplied from the interface or mixer through the mic cable. If you plug in a condenser mic and nothing happens, verify that phantom is on for that channel (mute the channel before turning phantom on/off).

Condensers generally sound more articulate, brighter, and detailed compared to dynamics. This doesn't always mean better, though, as it depends. They're quite sensitive, meaning they pick up room acoustics much more than dynamics, so that's why many podcasters stick with a dynamic.

Some condenser mics you might run into include the Shure Beta 87A, Audio Technica AT-4040, the Neumann TLM series, and Blue's Yeti mics.

Audio example 41: Neumann TLM103 condenser

Audio example 42: Audio-Technica AT2020 USB condenser

Audio example 43: Shure Beta 87A condenser

Audio example 44: Telefunken M82 dynamic

Audio example 45: Shure SM7B dynamic

Audio example 46: Electro-Voice RE20 dynamic

Audio signal levels

There are different strata of signal levels in an audio system, from AC power (highest) down to microphones (lowest). Most of the equipment operates somewhere in the middle, known as *line level*. Since the output signal of a microphone is significantly lower than this, we have to run it through a special amplifier that increases it to line level. These are called *microphone preamplifiers*. Due to the large amount of signal gain required, the amp has a significant impact on the sound, and different models are designed (mostly intentionally, sometimes not) to introduce a variety of flavors. Recording studios usually have an array of mic preamps so engineers can select favorites for each miking situation. Quality does count, however, and cheap preamps will sound, well, cheap. They don't provide enough gain to power mics adequately, they sound brittle, thin, and, well, cheap. As much as you can, splurge on the mic and preamp and it'll make a big difference in your show quality.

Directionality

There are times when sound needs to be picked up from only one direction, other times from all around the microphone. Mics are designed to be selective from where they collect sound waves. We call these directional characteristics *polar patterns*, and there are three primary types:

- Uni-directional (cardioid)
- Bi-directional
- Omni-directional

A cardioid microphone will collect sound primarily from on-axis, meaning the front of the microphone. This is useful when you don't want a particular mic to pick up other sounds that are intended for other mics. However, they don't completely disregard sounds from off-axis. As you get farther away from on-axis (moving toward the side and back of the mic), the microphone gradually attenuates the sound it picks up. This results not only in a signal level reduction, but also causes changes in the frequency response of the sound, altering its tone. This means that the sound will be unnatural as you get more off-axis; the term is *off-axis coloration*, and it's something we try to avoid. What this means for podcasters is that if there is a group of people at a table, each with their own mic, the person sitting next to another will still be heard in that other person's mic. And it won't sound as good or natural due to the off-axis coloration (it'll also sound more roomy and reverberant due to the increased distance).

Bi-directional microphones pick up sound from in front and in back of the mic, but reject sounds from each side. Omnidirectional mics accept sound waves from all around, though there is a slight directional characteristic for high frequencies, resulting in a slight roll-off from off-axis of the microphone. Omni setups can be useful for a group of folks at the table with only a single mic.

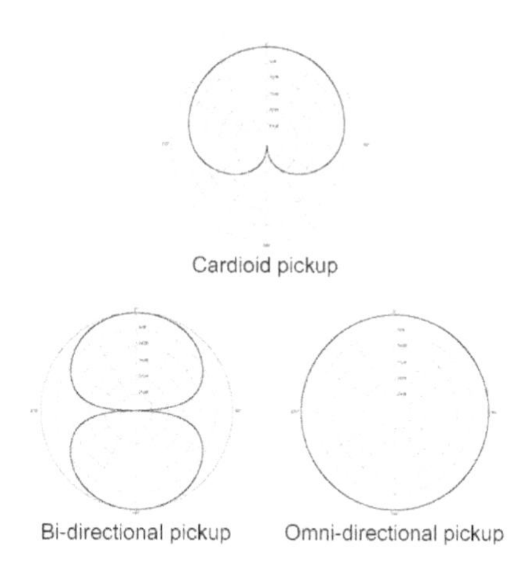

To achieve directionality, microphones have ports (openings or slots) along the outside of the casing. These ports lead to internal chambers that are designed to delay sound waves before striking the diaphragm from the inside. Cancellation is achieved when a soundwave coming from behind the microphone enters these ports while also diffracting (bending) around to the front of the mic to strike the diaphragm from the outside. When these two identical sound waves hit the diaphragm at the same time from opposite directions, the result is cancellation. Back to physics class, if two people push against each other with the same force, the result will be...zero. Nobody moves. Now, with audio signals it won't be total cancellation, depending on relative position to the mic and the fact that all frequencies have different wavelengths. The idea is that different frequencies will be attenuated or cancelled as the source moves around the microphone, resulting in a colored, altered sound. The lesson here? Listen carefully to each of your mics to hear what they're picking up; you definitely don't want to discover this when mixing the show and it's too late to fix it. You might need to space your participants away from each other a bit more, face them toward each other so as to present the null-side of the mics, or even better, build a podcasting studio with separate booths.

Shotgun microphones take the off-axis cancellation approach to the extreme. On-axis sounds enter with little attenuation, but sounds entering progressively off-axis are severely attenuated, particularly higher frequencies. These mics are intended for very tightly focused aiming, such as actors filming on set (or eavesdropping across the football field).

A common alternative design for implementing polar patterns involves dual capsules (diaphragms), one being omni-directional and the other bi-directional. These two elements are mathematically summed together as needed to achieve any of the three primary patterns as well as variations on these (hyper-cardioid, super-cardioid, etc). These microphones will have a switch for selecting the desired pattern—make sure you pay attention to which one it's set to.

Frequency response

As audio passes through any device or system, the relative balance of low, mid, and high frequencies is affected. Ideally we usually want a device to keep the original signal intact with no change. Microphones are a different story, however. I mentioned a few moments ago that microphones are intentionally designed to produce unique sound nuances. Microphones with large diaphragms tend to reproduce low frequencies better, making them good candidates for kick drums, upright bass, radio DJs, and podcasters. Condenser mics handle high frequencies better, so they can be ideal for cymbals and detailed, articulate vocals. Some microphones feature very flat response and are intended to remain neutral to the sound. Others, like the SM57 and 58, have a frequency response bump that adds energy in certain regions. There are no rules, but these characteristics can provide some guidance when selecting microphones. When you buy a mic, it comes with a frequency response chart to show how it passes sound, so check it out and compare with other mics to see how they respond differently.

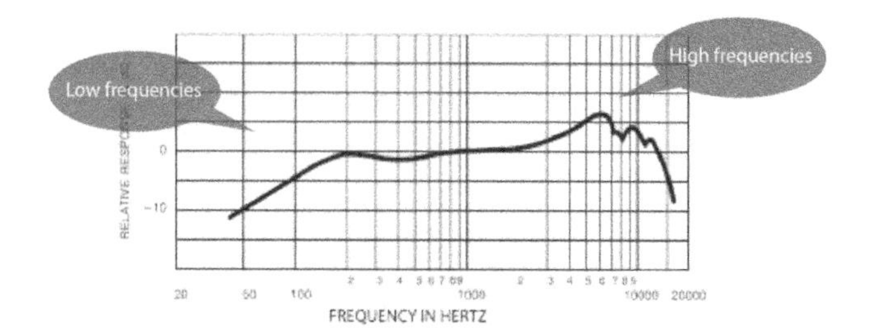

Something to be aware of is that low frequencies are over-emphasized when using cardioid or bi-directional microphones very close to a sound source. This low-end boost is known as *proximity effect.* Sometimes this is desirable, such as for giving radio DJs that characteristic deep, booming voice. For music it mainly muddies the sound and should be reduced as much as possible. There are a few ways to accomplish this, such as moving the mic away a bit, using an omnidirectional mic, or by switching in a low-cut filter. Many microphones have a switch that attenuates low frequencies, usually around 75Hz or 80Hz, and EQ plug-ins provide variable frequency select, which is great for tuning it to a particular voice range (James Earl Jones vs Minnie Mouse).

Transient response

How fast and accurately does the diaphragm respond when an acoustic wave strikes it? It takes time for the diaphragm to move and for the electrical signal to be generated. Audio transients are very brief bursts of signal found in all sounds, especially percussive events such as drums, guitar strumming, even piano chords. They are often so short many microphones simply ignore them. High frequencies have very short wavelengths and fluctuate very quickly. The quicker the response of the microphone, the more accurate the reproduction. Condenser microphones exhibit a faster, more accurate response that reproduces clear high frequencies, whereas dynamic mics, particularly ribbons, are more sluggish and tend to round off the waveform, resulting in a smoother, more mellow sound. Neither of these is better than the other—they're just different. So, if you talk into a condenser mic and it sounds tinny and bright, replace it with a dynamic and enjoy the mellower sound.

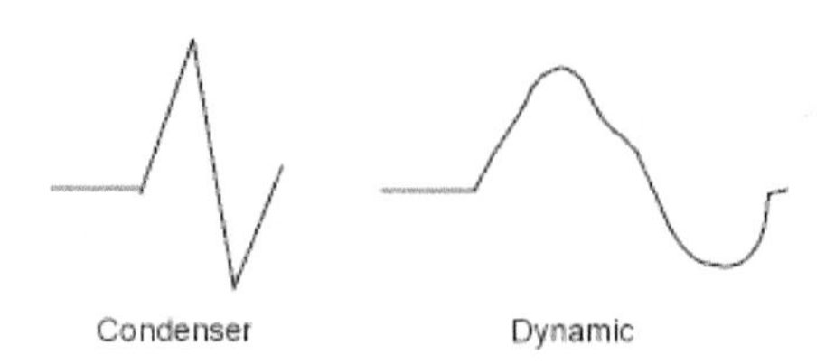

An exaggerated comparison of transient response

Sensitivity

This specification indicates a particular microphone's capability for output voltage level. A higher rating provides more output signal, which means you don't have to turn up the mic preamplifier on the console as much. This is beneficial because less gain means less noise and a cleaner signal. You'll notice this particularly when comparing condensers, dynamics and especially ribbons, all of which exhibit very different output levels (condensers are hotter primarily due to their internal preamp, but the idea remains the same).

Overload

Just like with any audio device, there is a limit of how much signal level a microphone can handle. A mic is overloaded when the SPL level (sound pressure level) is so high it distorts (clips) the diaphragm and/or electronics. High level sources such as drums or brass can potentially cause problems, though most microphones are pretty robust. Dynamic mics feature a higher overload tolerance than do condensers. If you're distorting the mic itself (as opposed to overloading the mic input on the interface), either move the mic farther away or use an attenuation pad if the mic has one. I once recorded a vocalist who had such a powerful voice she was overloading the mic—and it was a very nice, very expensive condenser. The mic pad did the trick and we got a great sound on that album. For podcasting I doubt you'll ever have a problem with mic overload; the main issue for podcasters is overloading the preamp and distorting the signal. That's an easy fix—turn it down.

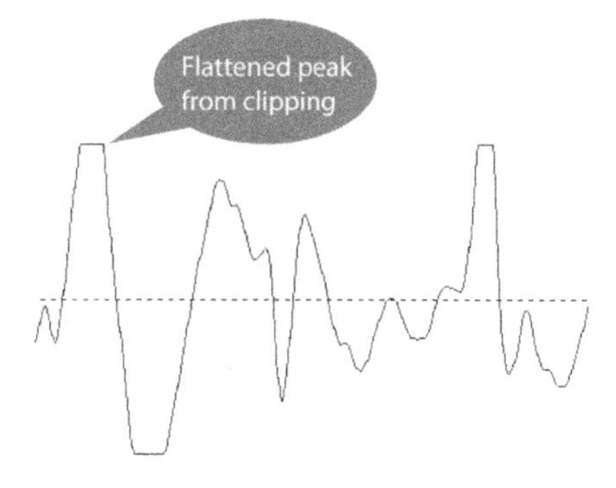

The waveform gets clipped when the device can't handle the high signal level. This actually changes the frequency components in the sound, resulting in a different tone we call distortion.

Impedance

This little-understood term is actually quite important in the design and connections for audio equipment. For now we'll just leave it as the mic's ability to provide a certain signal "strength" as compared to what the audio interface is asking for in order to obtain optimum transfer of signal. Remember when you go to a concert and they have restricted gates and entrances to control the crowd going in? Think about if they either closed these nearly shut or opened everything completely. People would pile up trying to get in or you'd get a stampede. You get the idea. Maybe.

In more practical terms, all professional microphones (and professional audio equipment in general) are low impedance (low-Ω), so don't use high-Ω mics, which usually have 1/4" connectors and act as radio antennas for any available broadcast that happens to be floating through the room. Low-Ω mics are much better at preventing outside interference and extraneous noise (motors, fluorescent lights, radios). High-Ω mics also suffer from high-frequency loss over distance.

Balanced microphone cables

Professional low-impedance (low-Ω) microphones use cables employing two signal-carrying wires in addition to a ground wire (shield). These signal wires are twisted around each other throughout the cable, and the shield is most often braided around the two wires. This provides maximum protection from outside noise interference, or RF (radio frequency).

How does it do this? Audio signals are AC current, meaning they alternate positive/negative between the two signal wires. Any outside interference leaks into the cable as a common polarity DC signal. When it arrives at the end it's cancelled out because all balanced audio gear (professional equipment) is designed to accept AC signals only. The shield in the cable drains extraneous noise by shunting it to ground.

Balanced audio requires a connector with three points, so we use either XLR connectors or TRS 1/4" (tip, ring, sleeve), and the three pins are numbered so as to match on each end of the cable. Pin 1 is always the shield (ground); pins 2 and 3 carry the alternating polarity audio signal.

By the way, none of this applies to USB mics as it's just digital data coming out of the mic.

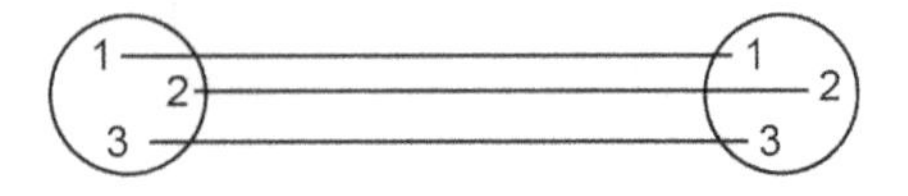

Each XLR connector has three pins for the three wires in a mic cable.

PZM (Pressure Zone Microphone)

The generic term for *PZM* (a product trademark of Crown, Inc.) is *boundary microphone* and refers to a microphone where the mic element (diaphragm assembly) is mounted on a flat plate. The concept is to reduce phase cancellations that occur when using a traditional microphone stand. A mic on a stand is elevated above floor level by several feet. The sound reaches the mic, but also bounces off the floor and into the mic somewhat later than the original wave. Back to acoustics, when two identical waveforms arrive at different times, phase cancellation occurs, which means the tone of the sound is altered in a negative way. With a boundary mic, there is no phase-altering reflection from the surface since the mic is directly mounted on that surface; all it gets are direct soundwaves from the source.

So, why would a podcaster care about these? If you're recording a group discussion and don't have individual mics, you need an omnidirectional microphone set in the center of everyone. This could be an omni on a short stand, but a PZM would be ideal. Just set it in the middle of the table, preferably on a rubber mat (or something sorta squishy), and it'll work fine.

Tube microphones

I mentioned earlier that condenser microphones require power to amplify the internal low-level signal. This amplifier can be either a FET amp (solid state) or a small vacuum tube. As in all things audio, tubes are desirable as they introduce color to a sound, specifically adding third-harmonic distortion into the signal path. This translates into a warmth and smoothness that engineers and musicians really like, so there are many tube microphones on the market these days.

Microphone switches

Pay attention to the various switches on your mics and make sure they're set to what you want. Polar patterns are typically set to cardioid (front pickup only), and filters and attenuation switches are off until needed. Not all microphones have these switches, so if you don't see a polar pattern switch on your SM57, don't tear it apart looking for one.

- Polar pattern select (condensers)
- Attenuation pad to reduce incoming signal level (condensers)
- Low-cut filter to attenuate low frequency sounds (condensers & dynamics)

Direct box

These devices connect electronic instruments such as keyboards or guitars to the console or audio interface. They're available in either passive or active models. An active DI requires phantom or battery power and uses active electronics to provide a "hotter", perhaps more aggressive output. You can find very nice passive models as well; the point is to buy good ones—the cheap models sound terrible. A decent DI can be had for $100 or so, such as from Radial or Nady, while the really nice ones go for several hundred dollars.

Direct boxes have a 1/4" input jack, where you plug in the guitar, and an XLR output jack where you connect a standard mic cable to the console or interface mic input. Once you plug everything in and turn on the channel, listen for a hum (ground loop). If you hear it, flip the ground lift switch on the box. It doesn't matter which setting you use, just pick the position that reduces the noise.

Some DIs also have an instrument/amp switch, which sets the expected incoming signal level. If you're plugging in a guitar or other instrument, set this to "instrument", which is lower. You can, however, take the output of a guitar or keyboard amplifier, which has a much higher signal level, and run that through the box. In this situation set the switch to "amp".

And in case you're wondering, "DI" stands for "direct injection", meaning the box takes your signal directly into the system rather than through a microphone. Nobody says "direct injection", though, so I wouldn't try it out on your friends. They'll just stare at you.

Direct boxes perform the following three functions:

- Reduces output level of the instrument (roughly line-level) to mic level so it can feed a mic preamplifier.
- Changes the instrument's high-Ω output (unbalanced line) to a low-Ω source (balanced) as needed for the mic input.
- Isolates audio signal, eliminating ground loop hum.

SEVEN
EQ AND FILTERS

We can examine sound in three dimensions: frequency, amplitude, and time. Each of these can be manipulated with signal processing, affecting a sound's timbre, volume envelope, and how it develops over time before fading away. The next three chapters explain the relevant processing for each domain and how they are typically applied.

Equalizer

An equalizer, or *EQ*, is a circuit that changes the frequency response of a sound by boosting or attenuating selected frequency bands. Think of it as a sophisticated tone control that allows you to make a sound brighter, less boomy, and so on. To understand how it works, you need to grasp the concept of what sounds are made of.

There is a wide range of frequencies that are audible to humans—anything vibrating between 20Hz and 20kHz falls within our hearing range. All sounds coming from musical instruments, voices, or traffic on the highway produce a very large number of frequencies—sine waves—which fall somewhere within that band. What makes a trumpet sound different from a fog horn is the difference in *which* specific frequencies and *how much* of each frequency is included in that sound. In other words, the harmonic content of that sound is what makes it unique. We call this the *timbre* of a sound.

With an EQ, we can actually increase or decrease regions of frequencies within a sound—this alters the harmonic structure and therefore makes it sound different. You're not turning a flute into a guitar, but you can make the flute brighter by boosting that sound's upper frequencies.

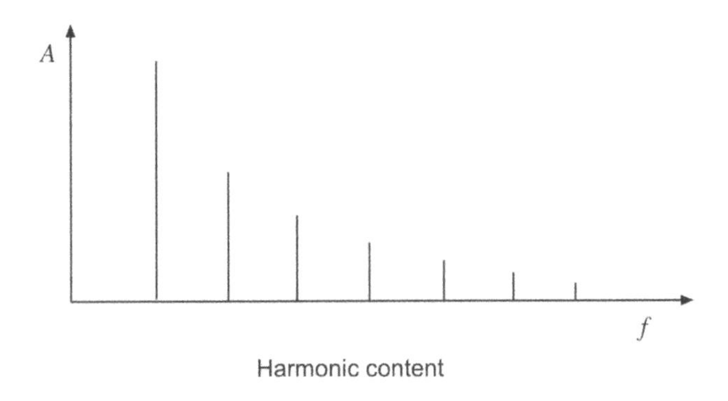

Harmonic content

Equalizer types

There are a few main EQ designs that give you control over a signal's harmonic structure, and each of these employs different types of *filters*, which are the circuits that actually do the work. First, let's introduce the EQs you'll run into, then I'll explain the filters they use.

Bass & Treble

The simplest EQ around, this type gives you control over the bass (low frequencies) and treble (high frequencies). That's it. This is the most common type found on home and car audio systems. They use shelving filters (explained a bit later) and function pretty much like the high and low frequency controls on your console or plug-in. They're based on the Baxandall EQ curve, developed in 1950 by Peter Baxandall in England. Although you won't find an EQ on your console called treble and bass, the underlying Baxandall curve is favored in many mastering studios for providing very broad, musical enhancements to a mix.

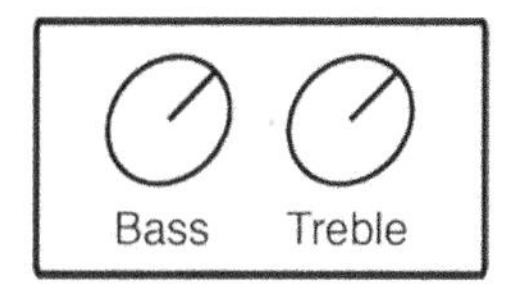

Graphic

Graphic EQs are easily identified by the row of vertical sliders that are used to boost and attenuate narrow frequency bands. These sliders are preset at certain intervals throughout the frequency spectrum, so you can only select what's there to change its level—you can't change the exact frequencies. These are found in some home audio systems, car stereos, as well as professional applications. They're sometimes used for "tuning a room" when an engineer adjusts the speaker system to compensate for acoustic issues in a particular room or performance hall. You determine which frequencies are causing problems, such as persistent feedback, then reach for that particular slider to turn it down. Most digital consoles provide graphic EQs on buses (aux sends, groups, main mix).

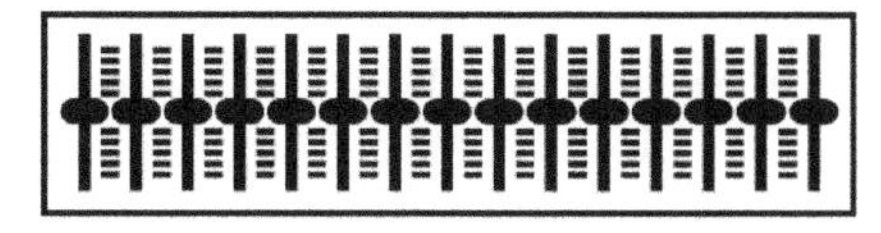

Graphic EQ

Parametric

Parametric EQs are the most common type found in plug-ins and mixing consoles. This one is the most complex, usually employing both peaking and shelving filters, and provides not only the option to boost or cut certain frequency ranges, but also to dial in the exact frequency region you want to work on. Whereas graphic EQs come preset for certain bands, parametrics allow you to move around and select where you want to work.

Parametric EQs divide the audio spectrum into a number of bands (regions), the most common being low, low-mid, high-mid, and high. Each band offers controls for boost/attenuation, frequency select, and perhaps bandwidth. Boost/attenuation is how you turn things up or down in that area, such as making it brighter by turning up the high frequency control. Frequency select allows you to slide up and down the spectrum searching for the exact area that sounds good or bad; once you've found it either turn it up (for more goodness) or down (for less badness). Bandwidth refers to how much of the sound is getting altered. Turn down a fairly wide chunk in the low-mid, for example, and it might sound too thin. You've cut too much body out, so narrow the bandwidth (also known as Q) and target the exact muddy region. We can also use a very narrow bandwidth to reduce a specific frequency problem, like a 60Hz hum.

On the diagram below, look at the section labeled "low-mid freq" and you'll see these three controls. In many EQs the mid-range bands use peaking filters while the high/low controls are shelving. Sometimes this is switchable between shelving and peaking—either look at the labeling or check the manual (gasp).

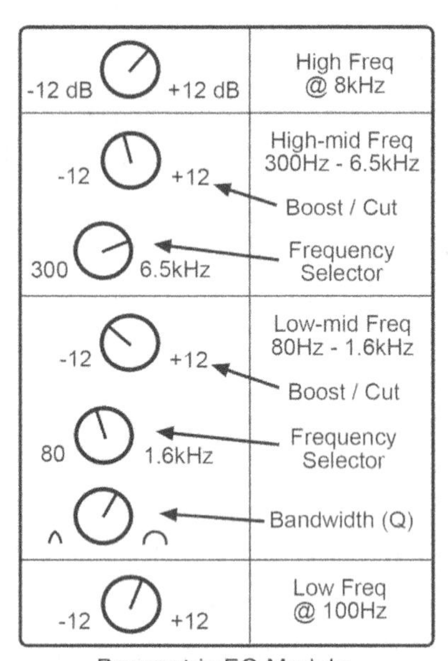

Parametric EQ Module

Parametrics give you very precise control over your sound, but it takes some understanding and practice to use them well.

Active dynamic control

Overlap of terminology happens too often in audio, and in this case there are two different concepts of the word "active" to distinguish here. Lots of different audio devices, including direct boxes, speakers, and processors, come in either passive or active designs. Passive models employ capacitors, inductors, and resistors to do their work, whereas active devices have transistors or tubes that provide amplification and control of electrical signals. There's nothing inherently negative with either approach as they feature different sounds and results. Engineers have been using active and passive EQs in the studio for decades. For example, the passive Pultec EQP-1A has long been famous for its huge bottom end, ultra-smooth response, and musical sound. (It has a tube on the output for amplification, but this is for color and make-up gain, not part of the filter circuits.) EQs with active components use transistors or tubes in the filter circuits, which color the sound a bit. But, the overall concept for both of these designs is the same—you're merely setting a fixed adjustment of frequency component amplitude in a sound.

Now, what we're talking about here are software EQs that feature active dynamic control over frequency processing, which is a fairly recent development. Active means it's continuously examining and responding to incoming audio; dynamic means the processor is affecting amplitude. A standard EQ will always boost or attenuate a certain frequency range to a set amount no matter what's happening with the audio signal. Wherever you set the slider or knob is how it will continuously affect the sound. An active EQ functions much like a combination EQ and compressor. You set the frequency regions where you want EQ to operate, but these adjustments only kick in when an amplitude threshold is met. Think of a vocalist singing through a song. Their voice timbre remains fairly consistent until that big climax in the bridge, when they really go for it and the voice becomes more strident or piercing. Set an active EQ to kick in at this point, reducing that particular frequency band, then letting go as soon as the voice calms down. This is a much more intelligent and musical design in that it makes a difference only when needed. I usually insert these on show host tracks as they gently shape dialog for more consistency.

Linear phase

Traditional EQ circuits introduce a certain amount of phase shift within the audio signal. As part of the filtering process, tiny timing adjustments occur as it attempts to rebalance frequency components within a region. Of course changes in timing affect the overall coherence of the signal (phase), so it further alters the timbre in a way that sometimes doesn't sound that great. Engineers have lived with this forever, so it's just part of life; there were times when I decided against using an EQ at all because the cure was worse than the problem I was trying to fix. On the other hand, this is partly how an EQ adds color to a signal, so that can be a good thing. Design engineers have worked to minimize this issue, resulting in what they generally term *minimum phase EQ*. This includes most of the EQs you'll find in your rack or plug-in collection.

Linear phase EQs eliminate phase issues by approaching things in a completely different way. They delay the signal so that changes can be processed, taking time to crunch the numbers, then output the result in a time-coherent signal. Processing is completely transparent with no coloring of the sound; it's generally used for fixing problems, such as surgically removing a narrow band. Is it better than traditional EQ? As always, it depends on the situation. Most of the time you'll want to use normal EQs on your tracks due to the sound they provide. Linear phase models are terrific for targeting problems; other than the lack of the traditional EQ sound, the main side effect is a fairly significant processing delay (latency), which prevents them from working well in real-time situations.

Filters

Filters are what actually do the work in an EQ. As we've described, different EQs employ one or more types of filters. Understanding the distinction will go a long way in helping you use the tools more effectively.

Peaking filters

Peaking filters allow you to boost or attenuate a range of frequency components centered around a specific point, called the *center frequency*. You can't operate on just one frequency, though—it always affects a certain number of frequencies on either side. This creates a bell-shaped curve and is known as *bandwidth*. The bandwidth of frequencies affected is called Q, for quality factor, and it can be widened or narrowed in many parametric EQs.

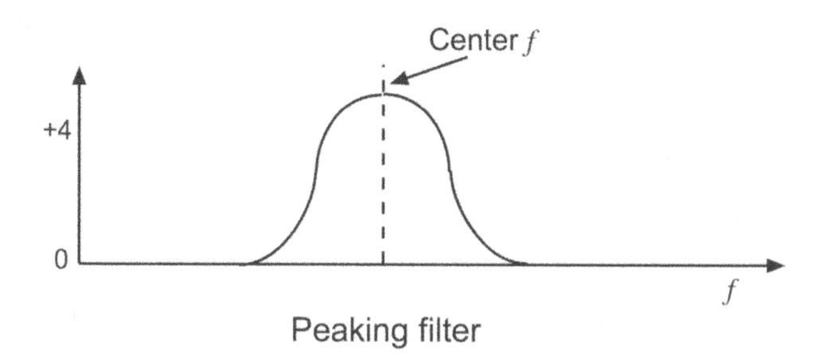

Peaking filter

The term Q is easier to use instead of precise frequency counts because the number of frequency components per musical octave doubles as you go up the scale. So, if you boost an octave in the low end, you might be adjusting forty frequency components, whereas in the high end that same octave could be working on eight thousand frequencies.

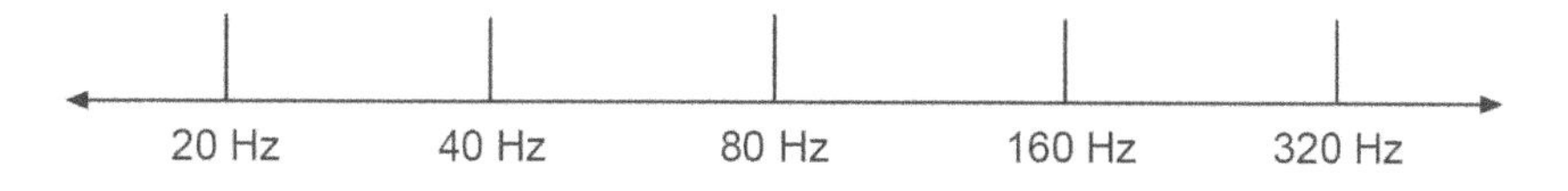

They both sound like musical octaves to us, so we need a system that compensates for the difference in actual frequency components. Q does this quite nicely by providing a simple number to refer to; a Q value of 0.8 sounds exactly the same regardless of where in the frequency spectrum you're working. Most EQs will label this control either with the numeric Q value or with a graphic that looks like a bell-curve.

All this means is that you can dial into any specific frequency location you want to fix or enhance. It might be a small slice, such as reducing a 60Hz hum, or it might be a wide chunk of your sound, such as boosting high-mid frequency articulation on your voice.

Use the following formula to calculate Q:

Q = Center frequency/Bandwidth

Example: If the center frequency is 1kHz and the filter is 1/3 octave wide, and its bandwidth lies between 2.5kHz and 3150Hz, then the bandwidth = 650Hz. Therefore, divide 1000Hz by 650Hz, resulting in a Q of 1.54.

- Larger Q values = narrow bandwidth
- Smaller Q values = wide bandwidth (affects larger range)

A couple more examples:

Center frequency = 1kHz
Bandwidth = 10Hz
1000 / 10 = 100 (large value)

Center frequency = 1000 (1kHz)
Bandwidth = 3000 (3kHz)
1000 / 3000 = .333 (small value)

Shelving filter

Shelving filters are often used for high and low frequency controls. They provide a boost or cut at a selected frequency (*turnover frequency*) at which point the adjustment remains constant throughout the rest of the spectrum. So, while a peaking filter will affect a range of frequencies and taper off on either side of your selected band, the shelving filter affects everything equally from the turnover point and beyond.

On the diagram below, the vertical axis represents amplitude (signal level) and the horizontal value is frequency range from low to high. Everything to the right of the turnover point is boosted. Everything to the left of this remains at 0 (unity), meaning the original level passes through unchanged. Note that the actual turnover frequency is 3dB down from maximum boost or attenuation, with the shelf flattening out beyond that point.

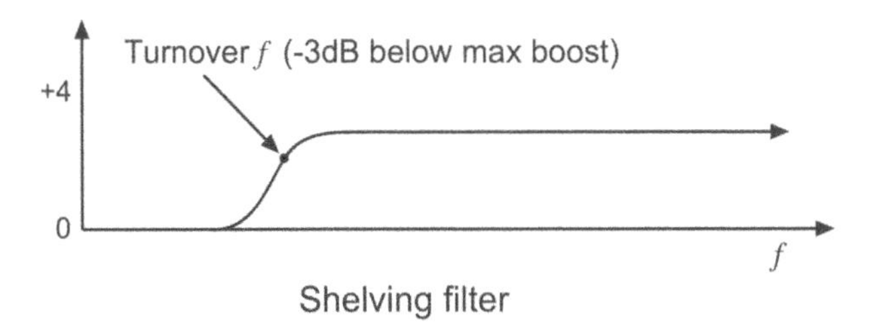

Shelving filter

Notch filter

Notch filters are peaking filters with an extremely narrow Q. They are designed to cut deep into a sound to eliminate a specific, narrow frequency range. One common application is getting rid of a 60Hz hum from a ground loop. I've also used these to notch out ringing overtones from cymbals and even guitar amps. The idea is to surgically remove an offending sound without interfering with the rest of the source if it can be helped.

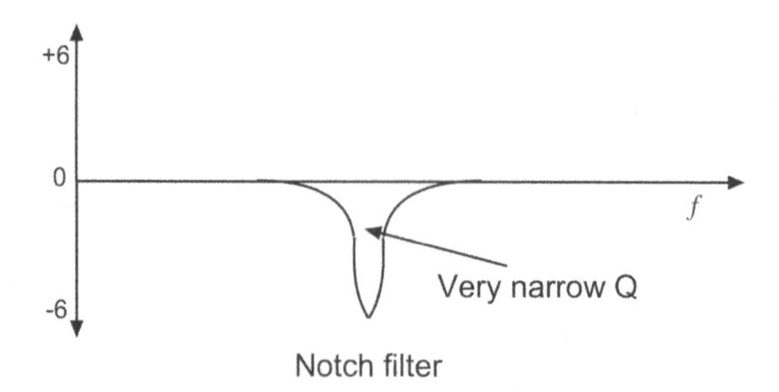

Notch filter

High- & low-pass filter

These work differently in that they provide no boost function. All they do is sharply attenuate frequencies above or below a certain point in the frequency spectrum. A high-pass filter set at 100Hz (*cut-off frequency*) will attenuate all frequencies below 100Hz, and a low-pass filter set at 8kHz will attenuate everything above this point. The rate of attenuation beyond the cut-off frequency is called the *slope*, which can vary between 6dB/oct, 12dB/oct, 18dB/oct, or 24dB/oct. Thus for every octave beyond the cut-off frequency the signal drops 6, 12, 18, or 24dB; higher numbers result in a quicker attenuation. There is no way to design a perfect "brickwall filter" that magically slices everything at the cut-off frequency. Actually, the cut-off frequency is about 3dB down from unity gain, meaning the attenuation begins slightly before this frequency.

Low-cut filters (also called high-pass) are great for getting rid of room rumble, low-frequency leakage from other sounds in the room, ground loop hum, and that annoying pop from a vocal getting too close to the mic. None of these are helpful, so get rid of it. Set them around 80-100Hz for voice unless you're somehow able to channel an interview with Barry White. The result is a cleaner sound without the low-end mud.

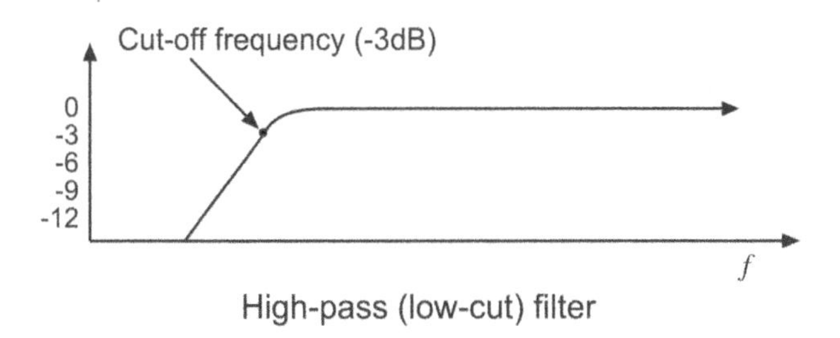

High-pass (low-cut) filter

Same, but different

Manufacturers have developed different approaches to the task of EQ for a reason—each one brings a unique flavor, advantage, and result to the table. Just because two models feature similar-looking parametric peaking filters in the mids doesn't mean they sound the same. Take a look at the two diagrams below.

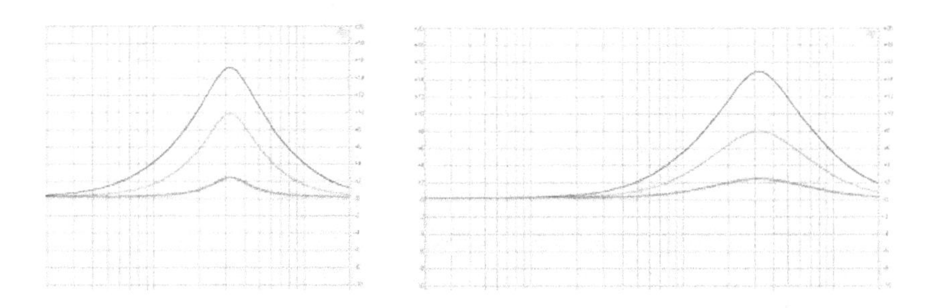

The left diagram features sharper curves and peaks, focusing closer around a particular center frequency. The EQ on the right provides a gentler, smoother curve and is generally considered more musical. Both are excellent designs, but they operate a bit differently and give you different results. Once you understand this concept, use your ears to select which type works for any particular situation.

Application

EQs are used to address tonal problems such as mic placement coloration, the physical setup of the instrument, or peculiarities of a voice. We can also creatively change the sound of a source to fit our mix better. If the show host sounds muddy, then attenuating frequencies in the low-mids can clean up the sound. For more presence and articulation, boost a little in the high end with a shelving filter. If you need to make a voice sound fuller with more bass, then boosting lows can help to a certain extent. But, if there aren't any low frequencies to begin with, then it does no good to try boosting something that doesn't exist. I once had an artist who kept demanding more bass on the vocal—and she was a soprano!

EQ should not be expected to compensate for poor quality sound sources or bad miking technique. Make sure everyone who's recording into a mic for a show is sitting in the right spot with the mic situated just so. It's also far better to find the right mic for each voice rather than trying to salvage a weird sound later.

Don't go crazy with the curves. It's always amusing to watch my engineering students start playing with EQs—it looks more like a work of art with towering mountains and lowly valleys among the bands. Be gentle with minimal boost or cut here and there. If a track seems to need 10dB gain or attenuation somewhere then it wasn't recorded well to begin with. It's a bit different for music, where it seems anything goes depending on your approach. There's a great Elton John record that was mixed with no EQ (no other processing, as a matter of fact) other than a tiny bit of high frequency boost on the drum overheads. Then there are other engineers who crank the knobs as far as they can go. Whatever floats your boat, but the difference is that professionals understand what they're doing and how it affects things. Learn the underlying concepts, develop your listening skills, then trust your ears.

A few remaining points to remember

Many engineers try to attenuate first; humans don't hear a cut as easily as a boost, so this is a more transparent approach to altering a sound. For example, you can brighten a track by attenuating low-mid frequencies. This can clean up the overall sound, allowing the upper-mids to shine through with greater presence.

Boosting frequency regions with EQ adds to the signal's overall dynamic range and level. Go too far and it will distort the audio signal chain. Always keep an eye on your meters.

High-pass filters reduce low-frequency leakage and noise, dramatically cleaning up your entire mix.

Some EQs are better at focusing on specific problems, others more for tonal shaping. I often use two on a track, the first to fix any issues and the second later in the processing chain to provide color in the mix.

Different EQ models bring a unique flavor to your tracks, so experiment.

Let's hear what they sound like

These first examples indicate the EQ setting while you listen. The settings are fairly extreme in order to make them easier to hear, and we'll use a music file for the same reason.

Audio example 47: Flat EQ setting (no change)

Audio example 48: 9dB boost at 1kHz

Audio example 49: 9dB attenuation at 1kHz

Audio example 50: 9dB boost at 4kHz

Audio example 51: 9dB attenuation at 4kHz

Audio example 52: 9dB boost at 8kHz

Audio example 53: 9dB attenuation at 8kHz

Audio example 54: 9dB boost at 125Hz

Audio example 55: 9dB attenuation at 125Hz

If your EQ provides bandwidth control, you can choose to adjust a big chunk of the sound or just a small slice. This next example features no

EQ, EQ cut with a pretty wide bandwidth, and then EQ cut with a much narrower bandwidth.

Audio example 56: Low-mid attenuation—flat, wide Q, narrow Q

Here's the difference between a high frequency peaking and shelving EQ. The peaking filter has a slight boost in the mid-range—it makes the guitar a little edgier. Then we switch to shelving, maintaining the same 3kHz frequency point. The difference is that it boosts everything above this point, so you hear all the noise and other nasties way up there. In this case we would prefer the peaking EQ, but it depends on the situation. Drum overheads, for example, usually work well with shelving, as do vocals, pianos, and so on.

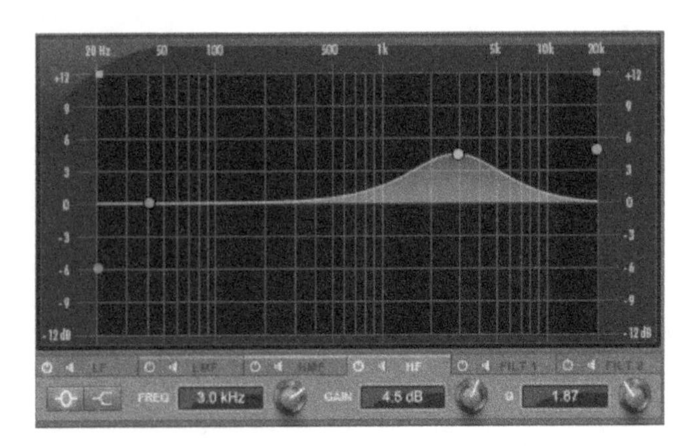

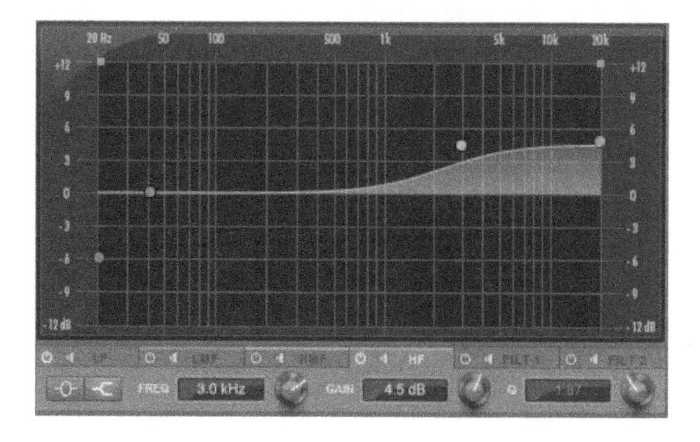

Audio example 57: High-freq peaking, high-freq shelving

Here's a voice example from my show. It starts bypassed (no processing), then with EQ on. The settings are a high-pass filter at 75Hz, narrow-Q 3dB boost at 160Hz, a narrow-Q -3dB low-mid dip at 340Hz, and a wide-Q 2.5dB high-mid boost at 3.7kHz for more presence and articulation.

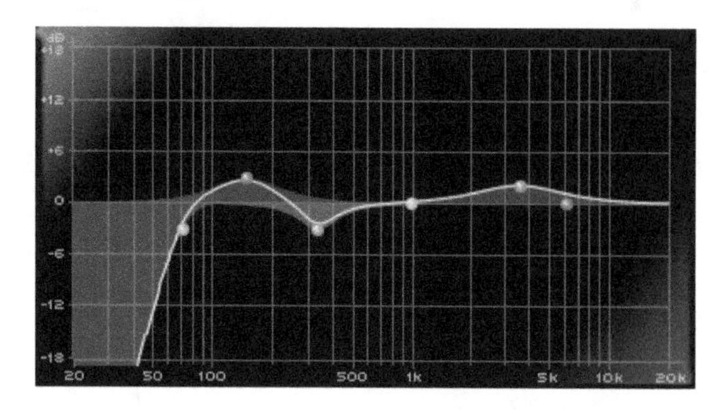

Audio example 58: Original voice track, then with EQ

For dialog, keep the following in mind:

- Mud and cloudiness often need attenuation. somewhere between 250 & 400Hz (varies, of course).
- Increase intelligibility, presence, and articulation around 4kHz.
- Avoid sibilance in the 6-7kHz range (don't boost here unless it's a really clean recording).
- "Air" and overall articulation and detail can be added with slight high-frequency shelving boost around 10kHz or so. The shelf actually starts below this point, so listen carefully as you set the turnover frequency.

Let's not forget your show music. Assuming you have a fantastic track that you paid for (keep it legal) or produced yourself, we can add just a bit more pizzazz to it with a simple, gentle low-frequency bump and high-frequency shelf. We'll start bypassed, then with EQ.

Audio example 59: Original music file, then with slight EQ

How to set an EQ

Some EQ adjustments are pretty straightforward. Need to make a cymbal or tamborine brighter? Turn up the high frequency control. For some EQs the high and low ranges only have boost/cut controls at a fixed frequency, such as 8k or 100Hz.

For the mids, which generally have more controls to work with, many engineers find that it's easiest to over-boost EQ in the range they're targeting, search for the exact frequency area that needs work, then make the final boost or cut adjustment. If a vocal needs a bit more presence, turn up the high-mid frequency gain control several dB so you can clearly hear the difference. Now turn the frequency select control back and forth to move through the frequencies. Listen for a spot that sounds good for what you need, then pull back the boost to a good level.

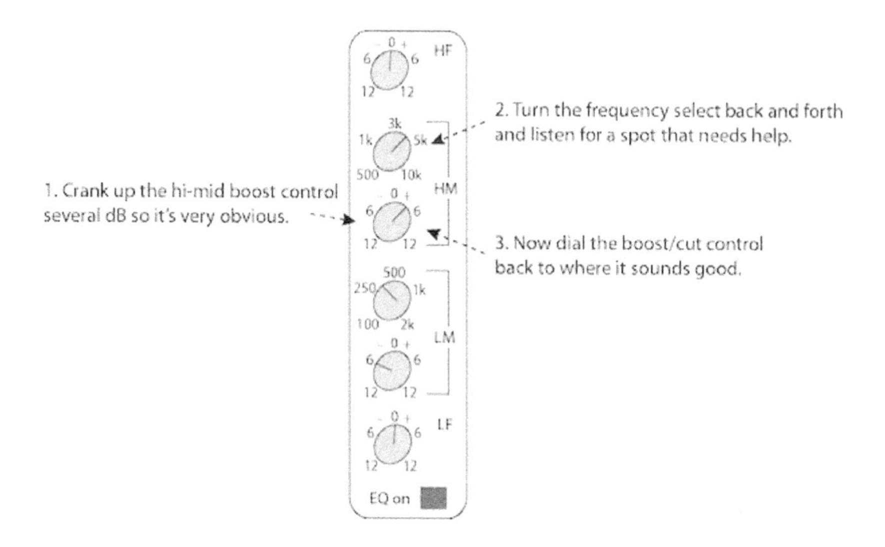

For this next audio example, we've looped a guitar part four times. First we boost low-mid EQ a lot so we can hear it, then sweep up and down looking for a muddy section. Hear how it gets really boomy and muddy near the end of this first loop? We pull it down below zero to attenuate it 3 or 4dB. The second loop starts bypassed so you hear the original sound, then we cut in the low-mid EQ band around the halfway point. Now for the third loop we do the same for the hi-mid band, looking for some nice natural presence and detail. I like the sound of the picking and strings just under 4kHz, so we reduce the boost so it's up about 2 or 3dB. The last loop starts completely bypassed, then we kick in the entire EQ. It might sound a bit thin and bright by itself, but depending on the song it should fit in nicely and cut through the mix.

Audio example 60: EQing an acoustic guitar

This process takes a lot of practice to hear what's going on and understand how the controls affect it. Over time you'll be able to hear what's "normal" and what's "not what you want".

EIGHT
COMPRESSORS AND GATES

THE SECOND OF the three domains we can examine in an audio signal is *amplitude*. This is the difference between a zero reference point and maximum electrical voltage or displacement of a vibrating object. A guitar string just before it's plucked is at a state of equilibrium, or zero amplitude. When plucked, the string extends beyond this point, and the harder it's pulled the farther it vibrates—making it louder to our ears. Electrical signals going through audio devices are the same thing, where amplitude is based on voltage level.

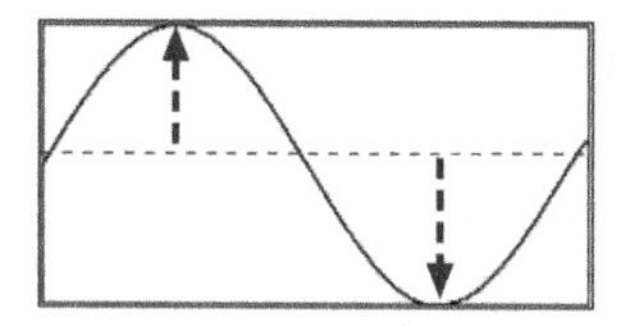

Amplitude of a signal

All components of an audio system, including the microphone, console, recorder, monitor speakers, and even your hearing, have varying capability for handling signal amplitude from the lowest level to highest peak. This is referred to as *dynamic range* and is indicated as a *signal-to-*

noise ratio (S/N). The level of a signal is measured between the noise floor of that particular component and the maximum level achieved just before the system goes into distortion. Modern equipment provides sufficient dynamic range as long as proper levels are maintained throughout the signal chain; it is the recording medium that has typically been the primary issue over the years. Professional analog tape recorders achieve an S/N of around 60—70dB, whereas the compact disc extends this to 100dB; modern 24-bit systems far surpass even this at 144dB S/N. An orchestra performance might possess a dynamic range of 65dB, but during low passages the noise floor of analog tape becomes noticeable. There is no noise floor in digital recording of any consequence, so the extended dynamic range allows us to record a wide-ranging performance without fear of distortion or noise intrusion (assuming recording levels are carefully set and monitored).

There are times when control over a signal's dynamic range becomes necessary. Radio requires heavy limiting due to restrictions on broadcast dynamic range. Noise floors for analog tape, and especially the consumer compact cassette, are fairly high and thus require some constraints on a signal's dynamic range. However, these days the issue is not so much needing to control amplitude ranges due to equipment limitations as it is fitting everything into a mix while seeking musical and sonic creativity.

A musician's performance usually varies quite a bit between soft and loud notes. Even during a relatively consistent rhythm part, such as strumming a guitar, performance fluctuations can cause significant differences in level. The same goes for vocalists, who may sing the verses normally, then kick into high gear on that dramatic final chorus. The result is that instruments and vocals will jump in and out of the mix, so some control over dynamic range will help keep things in their place.

For podcasting, show host dialog volume varies more than you might think. People trail off at the ends of sentences, make contemplative statements, or explode in fits of laughter. If you do nothing with this your listeners will have a difficult time hearing everything while driving to work.

Sonically, altering the dynamic range of a signal provides interesting timbral results that add to an engineer's options for crafting a mix. Because of this, many different types of amplitude-based processors have been developed over the years. Changes in electrical components, circuit design, and other factors are what give each model its special flavor. You should invest time trying a variety of processors on a track to get a feel for how they sound. Sometimes engineers aren't even looking for gain reduction, but to just color the sound in a pleasing way. Let's take a look at what a compressor does and how to use it, then listen to what it can do for your tracks.

A *compressor* is a signal processor that limits how much an audio signal varies from soft to loud. As the incoming signal gets higher, the compressor will reduce this so it sounds more even. Think of it as a volume cruise control for audio.

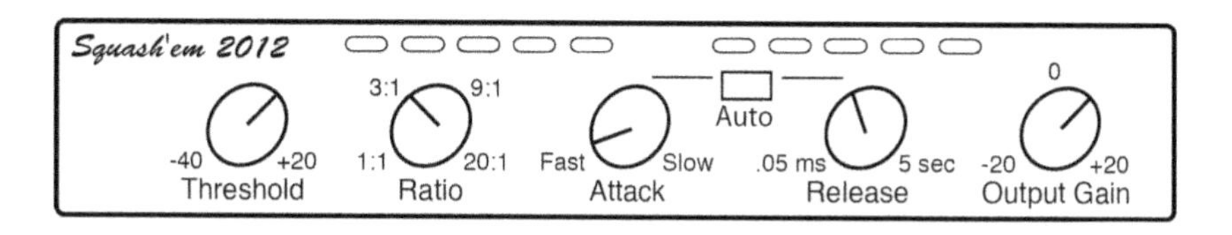

Compressor functions

Threshold is the signal level at which the device begins to compress. Any signal that goes higher than the threshold will get attenuated, meaning it won't be so loud coming out the other end. Threshold doesn't directly dictate how much compression will be applied, only the point at which the device starts doing its thing. A high threshold will only compress the highest, loudest levels in the signal, whereas a lower threshold affects most of the signal and is more noticeable.

Ratio sets how much the signal will be reduced. Once a signal goes over the threshold point, it will be attenuated at a rate set by the ratio. For example, if you set it at 3:1, every 3dB of signal beyond threshold will result in only 1dB coming out of the compressor (smaller dB numbers are quieter). Since these are ratios, multiples apply, meaning if 9dB of signal goes over threshold, it will output 3dB.

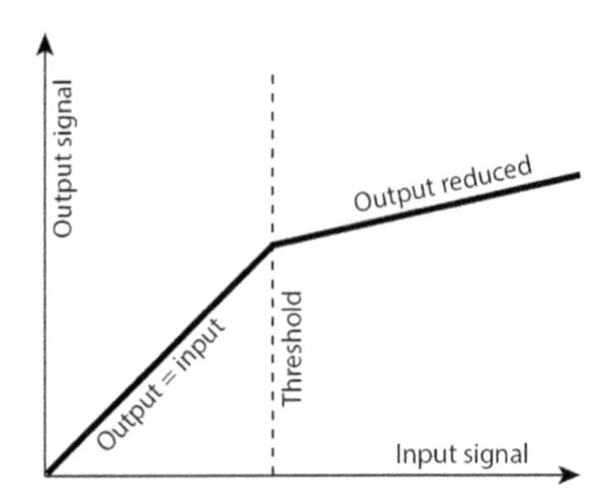

A signal going over threshold triggers the processor to start compressing, but we can control how quickly this happens. This is the *attack* time; a fast attack setting will very quickly trigger full compression once threshold is reached, whereas a slower attack will take its own sweet time to get there. This is a big deal because it affects the tone of the sound as well—a really fast attack will lose the initial bite or attack of the sound, giving you a more rounded tone. Lengthening the attack time will allow that bite to come through before it reaches the maximum compression ratio. The technical term for what you're shaping is called *transients*, which are the very first high frequency components of a sound that occur when an instrument is plucked or hit. So, turning the attack time up and down will give you a brighter or duller tone.

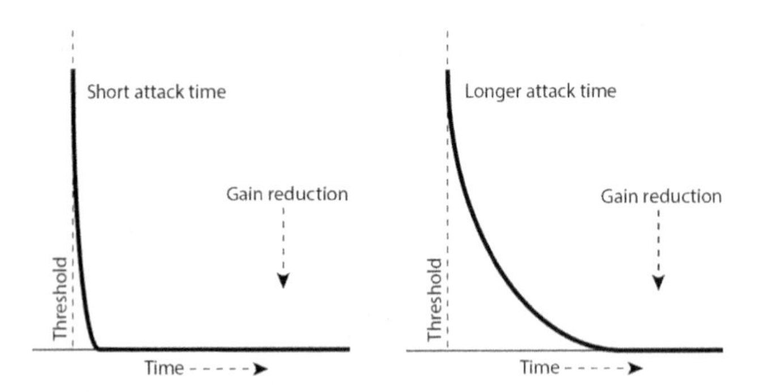

Many compressors also provide a *knee* setting, which effectively extends the threshold point into more of a range. Once the attack time triggers compression, the knee controls how gradually the compression will reach the ratio setting. A hard knee, indicated as a sharp bend, immediately alters gain from the 1:1 input level to the desired n:1 ratio setting. But a soft knee is more gradual, providing a smoother transition into full compression.

Once the incoming signal begins dying away and falls below threshold, the compressor will "let go" and allow the signal level to return to normal (unity gain). How long does this take? That's the *release* time you set. A medium release is usually pretty smooth, whereas a long release may prevent the unit from recovering before the next audio passage begins. For music this obviously depends on the tempo and what they're playing, whereas for dialog it could hold the track too low when the person isn't talking loudly. A long release could also extend the life of a musical note just a bit; as the original sound fades away naturally, the compressor is slowly returning to unity gain, briefly counteracting the decay. A very short release setting can cause the unit to cycle too quickly, resulting in a breathing or pumping sound. Most compressors feature an auto setting option for attack and release, and for the most part it works pretty well. As you get better at this, go ahead and fiddle with these yourself to make it fit each particular situation.

After a signal goes through a compressor, it's usually got a lower overall signal level and might sound quieter in your mix. To compensate for this, all compressors have an amplifier at the very end where you can crank it back up a bit (*gain*). Sounds contradictory, but what's happening is that once the compressor does its thing, you've got a more controlled signal with less dynamic level swings. Now you can take this streamlined signal and crank it up as needed. You set it in the mix and it won't jump in and out as much, meaning the show host will sound more consistent and easier to hear.

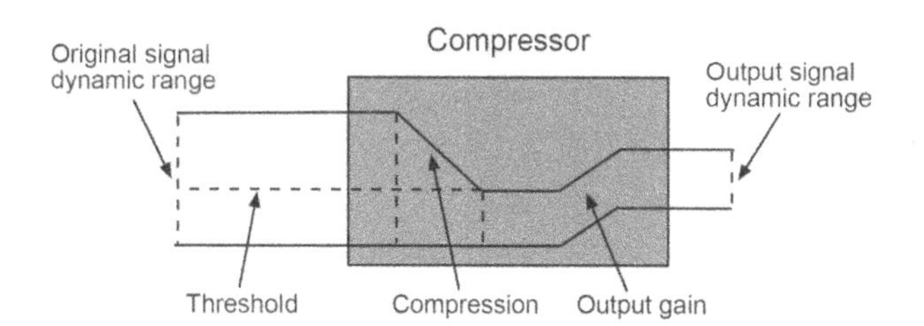

Compressor design and types

A variety of approaches have been developed for compressors, each with its own particular sound and results. Here are a few concepts you might run into.

VCA

VCA stands for voltage controlled amplifier, used in many different audio applications. In a compressor, the parameter settings (threshold, ratio, attack and release) are used by a control signal to tell the amplifier how to attenuate the audio. VCA-based units are excellent for fast-acting control; they react quickly to transients and peaks and are usually hard knee (with exceptions). They are not good for general smoothing or shaping of level and volume, but are quite nice for adding a punchy, aggressive sound. Most of your standard compressors fit into this category, so this is what you'll end up using for your show tracks (whether you realize it or not).

Bus

In the 1980s SSL introduced the G-Bus compressor, found on the mix bus of their large format recording consoles. Engineers mixed hundreds of albums on these boards, finding the compressor to be wonderfully transparent with the ability to "glue" a mix together. The idea of a bus compressor is to be subtle and bring a cohesiveness to the entire mix; you're only looking for a dB or two of gain reduction. They are typically VCA-based because this design provides a lot of control over the signal and works well on summing buses. Another popular model is the API 2500, which can be set for a more aggressive tone with lots of punch. Although the good hardware models are pricey, plug-ins are available based on these classics. If you're running a DAW, use the main stereo mix track to insert this compressor. I always put this on my overall podcast mixes, so consider giving it a whirl.

. . .

Optical

The optical compressor is so-called because it employs an electro-luminescent panel (light bulb) that glows brighter as incoming signal increases. A photo resistor detects this change and increases impedance, which then reduces signal gain. Decreasing input level will dim the panel, decreasing the resistor's impedance, thereby allowing the circuit to return to unity gain. There's something magical about this particular approach because it just sounds amazingly musical. The first unit of its kind, and undoubtably the most famous, was the Teletronix LA2A, utilizing an optical cell (T4) whose lineage derived from the Cold War missile programs. You can find them today as hardware reproductions from Universal Audio or as modeled plug-ins; they're particularly fantastic for music vocals, but perhaps not so suited for podcast dialog as their response time is rather slow.

Variable mu

These are tube compressors that feature an incredibly smooth, transparent, and warm sound. Slow to react, they shape a signal over time. An increase in input level will actually increase the ratio along the way. As opposed to other units that feature a tube for color, this design uses the tube itself to change gain. By the way, "mu" means "gain", in case you were wondering. These are popular for music mastering because they can provide a final, overall shape to a mix.

FET

Field-effect transistor-based compressors are the opposite of optical and variable mu in that they aren't even close to being transparent and smooth. These units aggressively add punch and color to a track, featuring a super-fast attack time. The Urei 1176 (Universal Audio) is probably the most famous and loved FET compressor ever made and sounds fantastic on snares, guitars, and so on. But probably not for podcasting!

. . .

Multiband

A standard compressor operates on a signal's entire frequency range. This can cause issues if, for example, there is a lot of low frequency information in the sound. This will cause the compressor to react more aggressively, thereby hitting the mids and highs too much. Multiband compressors divide the frequency spectrum into bands, or regions, such as low, low-mid, high-mid, and high. This concept should be familiar from the EQ chapter. Now the processor is free to examine each band independently, only applying compression as needed in a particular region. Such units have been utilized in the mastering process for decades and are also very popular for live reinforcement. They might be helpful for an overall podcast mix, but I wouldn't spend any time worrying about it. Play with one someday when you get a good feel for things and see what you think; otherwise just forget I mentioned it.

Dynamic EQ/compressor hybrid

Dynamic (active) EQs combine the real-time reaction of a compressor with the frequency-selectable precision of an EQ. Dial in the exact frequency regions and bandwidth that need tonal attention, but let the compressor determine when, and how much, should be applied. This is very beneficial in the studio, including podcast dialog (I use these, by the way), but can be particularly effective for live reinforcement.

Let's hear what they sound like

Listen to the following examples to hear what the compressor is doing to the track.

Audio example 61: Compression on voice (moderate, then too much)

Audio example 62: Compression on saxophone (none, heavy, medium)

Audio example 63: Compression on acoustic guitar (light, then heavy)

Probably the most noticeable result is the loss of the initial attack. The initial snare hit or guitar strum is the loudest part of the sound; if you have a fast attack setting on the compressor, it immediately holds this back. Listen how it takes a quick dip before leveling off a bit. Of course, different settings will change this somewhat, so put your headphones on and experiment by just turning knobs. Too much compression will make it sound squashed, but there are certainly times in music production where extreme settings sound really interesting.

Don't forget to try different types of compressors—each provides its own flavor. I typically use two compressors for each of the dialog tracks for my show. One is early in the chain, set to do most of the dynamic control, but doesn't color the sound. The signal then goes to an old analog-modeled plug-in that only kicks in a couple dB of reduction, but adds warmth and that great analog sound. Many podcasters don't do any of this, but I tend to like the result.

Notice that the meters will show either input, output, or gain reduction (GR). If the meter sits idle at 0VU, it's in GR mode (no reduction taking place); the indicator will swing negative once compression kicks in.

To get started, try these settings and then experiment a bit.

- *Ratio*: 3:1
- *Attack*: Keep it fairly fast, but not quite all the way.
- *Release*: Medium
- *Threshold*: Slowly turn it up or down until you get just a few dB flashing.
- *Output gain*: Start at 0 (no change) and adjust as needed to provide a good level in your mix. You may not need this if you're only compressing a few dB.

If the compressor meters light up like crazy, it'll sound squashed. You don't want this for normal dialog, so raise the threshold. It takes quite a bit of practice, listening, and experimentation to get a feel for setting these parameters, so get busy.

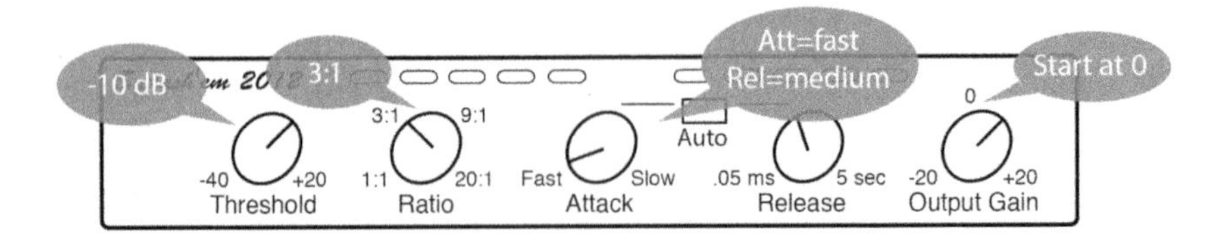

Limiter

Limiters are the same as compressors, only with a high compression ratio (10:1 or higher). The idea is that you want to aggressively control the signal once it goes beyond a certain point, so the threshold is usually high also. While they can be used just like a compressor on a track, they are also often inserted into sound systems to protect against sudden pops or transients that might damage something (like when someone plugs a microphone in without muting the channel).

Peak limiters are a type of limiter used during the mastering stage. The typical idea is to get as much volume as possible out of the mix so it sounds "just as loud" as the other record company's albums. It essentially squashes the signal, bringing low levels way up so it doesn't vary much dynamically. If applied too liberally you end up with a mix that's lost its musicality, ebb and flow, and tires the ears after listening for awhile; it'll also create massive distortion if pushed too far. There's a reason the knobs turn, however, and so we can set this at a more moderate level which can be quite useful. In chapter three we described an effective application of peak limiters for mastering your show files.

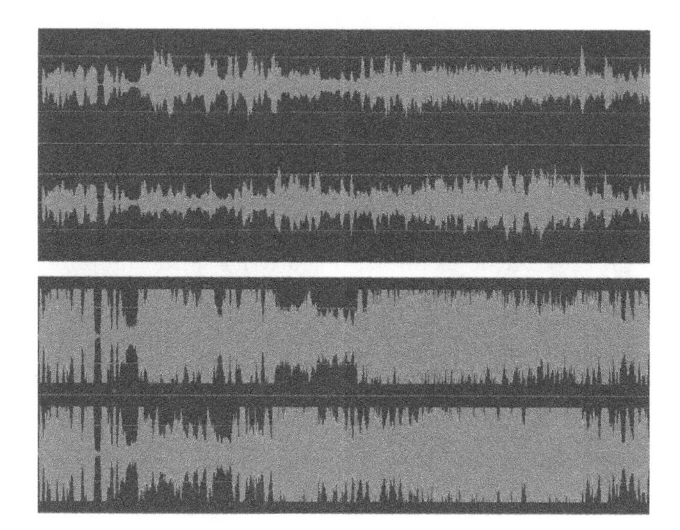

Original on top, squashed pulp on bottom

De-esser

With vocals or speech, the consonant "s" sometimes gets a bit out of control. We call this *sibilance*, and it's prominent around 6kHz or so. A de-esser plug-in reduces sibilance by applying compression whenever it detects excessive energy in that frequency range. But it doesn't compress the entire audio track, only the specific frequency band where sibilance occurs. Many de-essers provide control over exactly which frequency region it's targeting and how much compression you want it to perform. I always put a de-esser first thing in the plug-in chain if I hear any hint of sibilance.

Audio example 64: Sibilance

Noise gate

Microphones inevitably pick up sounds other than the one you're trying to focus on. It could be the other show host sitting across the table, the air conditioning vent overhead, or a ticking clock from the hallway.

Sometimes you can reduce this by inserting a *noise gate*, which is designed to attenuate a signal once it falls below threshold. Since these background noises are usually quieter than the main signal you're recording, you can set the gate to drop the signal when the main source is not active.

Let's take an easy music example, like a noisy guitar amplifier. Set the gate threshold low enough that it stays open when the guitar is playing, but will shut down (attenuate) while not playing. This will get rid of the amp noise until the music starts. You can't gate the noise when the musician is playing—the channel has to remain open so you can hear the guitar. However, the idea is that the guitar sound will mask most or all of the background noise, so it's not as big of a deal.

Gating a group of folks sitting around the table can be tricky, though, because it's really hard to find a threshold that will reduce leakage (the other people talking) without cutting out all those low-level "uh-huhs" and such. And if one person laughs loudly, every gate will open up because that'll certainly go over threshold. This is a similar problem to the "strip silence" approach we discussed in chapter three, where you set the DAW to completely erase anything below a certain threshold. This would remove all sound between the various phrases of dialog, but at the risk of inadvertently deleting quiet utterances that add to the discussion. Keep in mind, however, that the gate doesn't have to completely silence the track; it can be set to merely reduce signal level instead, which can improve the leakage situation without being too extreme. I dunno...I produced one episode using gates on everybody, but that was the last time. It took me forever listening through the show making sure I wasn't killing something important. Everyone has to develop their own workflow and preferences. It also depends on how fussy you are with the final show file and whether elegant mixing technique is worth the time.

By the way, another term for noise gate is *downward expander*, the idea being that the processor is increasing dynamic range by lowering low level signals. This is opposite of what a compressor does, and can be used not only for reducing unwanted sounds, but also to introduce more dynamic range to the sound.

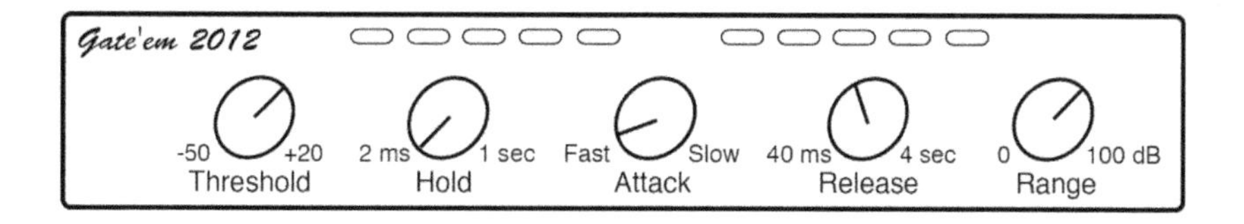

Functions and controls

Threshold is the point below which the gate will attenuate the channel. As the program material falls below this level the device reduces gain to a user-set *range* or *floor*. This may actually not be all the way down; sometimes it's sufficient to merely drop the signal a few dB rather than try to cut it out altogether. *Attack* and *release* times operate similar to a compressor. Attack determines how quickly the gate will open up once a signal comes in above threshold, while release sets how long it takes to "shut down". You need to play with these settings to find a combination that works for each specific situation; otherwise the gate might chop sounds off too abruptly or kick in too slowly. Some gates also provide a *hold* timer—after the gate has opened, it will hold it open until after this time has elapsed, even if the program material has already dropped below threshold. It's probably a good idea to leave this off while you're learning how to use the other parameters.

Audio example 65: Noise gate with abrupt settings, then smoother

Audio example 66: Guitar amplifier hiss with gate off, then on

---

NINE
REVERB AND EFFECTS

TIME IS the final domain we can control on an audio signal. Sounds have life cycles that take on various attributes over time before they fade away. The physical environment influences this significantly, and we can alter or recreate the time envelope of sounds in rooms with time-based processors. A basic foundation of sound behavior in acoustic spaces is necessary for understanding how these work.

For the most part podcast production doesn't involve any of these processing tools, such as reverberation or delay, but since it's weird only giving you two of the three dimensions of audio we should finish the story. Besides, maybe now you can think more seriously about cranking up that garage band you've dreamed about forever.

Sound generated in a space will spread out and reflect around the various surfaces. Size, shape, and surface treatments determine frequency response and the nature of these reflections, meaning they differentiate a bathroom from a concert hall.

A propagated sound passes through three stages:

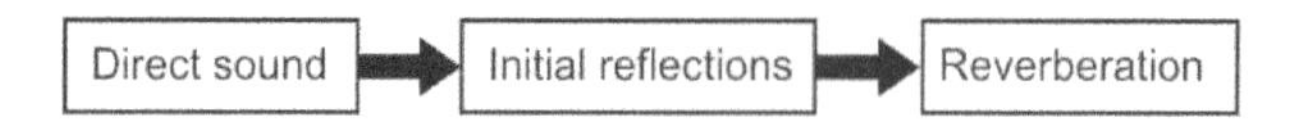

Direct sound is what you hear straight from the source, with no reflections from nearby surfaces affecting it. The closer you are to a sound source, the more direct sound you will hear.

Early reflections are the first waves to be reflected from nearby surfaces such as walls, ceiling, floor, or room furnishings. Unless it's a really big area, they will not be heard as distinct echoes, so you can't sit there and say "Oh, I just experienced an early reflection. That was pretty wild". Your brain, however, is able to detect these various waves and decodes that into a sense of how big the room is. This comes from the delay time between the original direct sound and these early reflections, so this is primarily how you "know" whether you're in an auditorium or small club. A longer initial delay indicates a larger room, because the reflective surface is farther away from the listener and takes a bit longer to arrive. If the delay between reflections is long enough, we hear distinct echoes. This occurs in very large, reflective rooms such as gyms as well as outdoor stadiums.

The reflections in a room rapidly multiply, particularly when the source continues to produce sound. As the reflections become more numerous and dense the result is *reverberation*. This evenly distributed, diffuse sound field contains no discernible reflections and will eventually die away at a rate dependent upon the physical attributes of the room (size, wall surface textures, and so on).

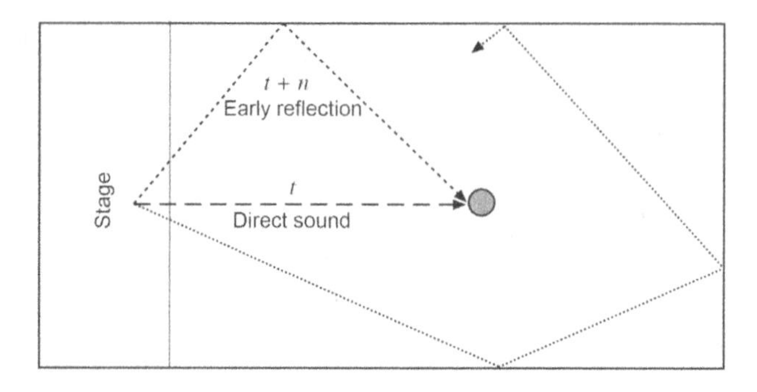

Time-based processors operate on these principles, allowing you to craft any environment you wish. Sometimes this is to merely recreate a familiar acoustic space for a performance, but many tools are used in creative ways to enhance your mix. Let's take a look at the more common processors, find out how they work, and hear what they sound like.

Reverberation

We always hear reverberation of some type and amount wherever we are. You've heard the difference between talking and singing in your living room, bathroom, and the school auditorium. Sometimes in studio recording we try to reduce reverberation in favor of a closer, drier signal. Then we can artificially add whatever we want for the final mix. Reverberation processors take incoming signals and generate numerous reflections, simulating any particular environment you want. Here are the main types of units we've seen in the studio over the years, all of which are now modeled in software.

Spring

Ever hear someone drop a guitar amplifier on the floor and it went "sproing"? That's an example of a spring reverb, where an actual spring is connected to transducers. Feed a signal into the unit and it travels back and forth through the spring, generating multiple reflections on the output. Dig into your old toy box and pull out a Slinky. Put your ear down at one end and tap the other to set it in motion—that's what we're talking about. These aren't used much anymore because they just don't work very well. I had to use a spring reverb many years ago, and every time the snare hit it overloaded the spring. Not very musical.

Plate

A plate reverb unit is a large metal plate suspended under high tension. You set it up in the room, sound bounces against it, and the vibration of the plate generates reflections that are picked up by transducers attached to the plate. These reflections are then fed back into the console. These also are not common anymore, especially since you can get much of the same effect from a digital processor (for a whole lot cheaper).

Chamber

Early studios had a separate room dedicated to generating reverberation. The engineer would feed a track to speakers mounted in a highly reverberant space; the sound would bounce around and be picked up by microphones that were then fed back to the console. This isn't done much these days because using a room exclusively for reverb is expensive and not really necessary due to digital processors that do such a great job. But a lot of great records were cut in studios using this approach.

Convolution reverb

This is a process where actual samples of a room are recorded, providing an acoustic profile of that particular space. An impulse response is generated in the room, which is a very brief, sharp transient burst (like a starter gun or bursting balloon). The resulting sound in the room is captured and stored as an impulse response model. Reverb is generated by combining an audio track with the samples from the model in a mathematical process referred to as convolving, where each sample of the audio track is multiplied with data from the impulse model.

Types of reverb patches (presets)

Halls

Hall patches simulate the environment of an auditorium or other large space. Longer first reflections, attenuated high frequencies, and denser, longer reverberation time are what make these sound distinctive.

Rooms / chambers

Smaller spaces are characterized by short first reflections and less reverberant sound. Think of a club or your living room.

Plates

A plate preset mimics the physical plate units described earlier. These typically sound brighter and crisper than hall or room settings.

Effects

Most processors not only generate reverb, but also produce lots of various effects. We'll describe a few of these a bit later.

User-set parameters

Pre-delay (early reflections)

After the direct sound occurs, there's a natural time delay before the early reflections start to show up. A larger room takes longer for these to arrive since the walls are farther away. Set the pre-delay time short for a small room, longer for a larger space. The range provided on effects processors is typically 0-100ms, but 0 is quite weird since there's no such physical space with absolutely no distance between the source and the nearest surface. Try between 10 and 30ms or so to give a vocal some space without losing it in reverb; going a bit longer can improve clarity of the voice as it provides more distinction between the direct sound (clarity) and the ambient environment.

. . .

Reverb time (RT60)

This represents how long it takes reverberation to die away to a point 60dB below its original level. A large hall will have a longer RT60 than your living room. When mixing music you want the RT60 to match the tempo and style of the song, so a faster song should have a shorter RT60 time; otherwise the reverb builds too much and muddies the mix. For slower ballads, a longer RT60 might be appropriate to allow the music to linger longer.

RT60 is frequency dependent, meaning high frequencies will attenuate much quicker than lows. It's also directly related to the room itself, including size and surface materials. Wallace Sabine, considered the father of architectural acoustics, developed an equation for determining the relationship between RT60, room size, and the overall absorption coefficient of the space. This enables studio and facility designers to predict how sound will behave in a particular room, even before it's built.

Hi/Lo EQ

Most reverb units provide a simple EQ control so you can brighten or darken the reverberant sound.

Wet/dry mix

The output of the processor can either be all effect (wet), all original (dry), or a combination. Since reverbs and effects are most commonly accessed via aux sends, we end up with the original, dry tracks and the reverb return feeding the mix bus separately. In this case the output for the reverb unit needs to be 100% wet. If a verb or effect is inserted directly on an audio track, then a balance needs to be set.

Other settings

Some simple reverb processors only provide a few settings, but others allow you to completely shape how the sound changes over time in a particular type of space. Once many years ago I had to transport a super high-end reverb unit that did just about anything you can imagine to another studio in my car, and for that brief period of time the value of my wheels more than doubled (hey, it wasn't *that* cheap of a car...the device cost about $10,000). Parameters such as size, shape, and diffusion fine-tune what the reverb sounds like, how long it lasts, exactly how it dies away, and so on. Play around with these various settings and see what you come up with.

How do you set it up in the DAW?

Reverb processors can be used for a single track, the band, or the entire mix. So, you could insert a reverb plug-in on a vocal track, making sure you set the output balance to 50/50 or so. But it's more powerful to keep the reverb on a separate track, giving you more flexibility in how to set it up.

Create a new *aux track* and name it something obvious like reverb. Insert a reverb plug-in on that track, and set the track input source to an internal stereo bus in the DAW. On the voice track, turn up an aux send and route that to the same stereo bus. Now you've got a copy of the voice audio going over to the aux track where reverb gets added to it. Now you can treat the reverb sound any way you want: EQ it, compress it, pan it, and so on. Need more overall reverb? Just turn up the fader on the aux track. Here's what this looks like, in this case adding reverb to several drum tracks.

Individual track sends combine to create a submix on the aux track, where reverb is added from the plug-in.

Application

Reverb is used for most any music recording project, but almost never for podcasting shows that are primarily dialog. Try different patches to see which is more appropriate to the situation. You will often use more than one processor and patch setting in each song, such as putting a plate on the vocal and a large hall on the drums. Also carefully listen for reverb time; make sure the decay of the reverb matches the tempo and style of the song. Don't just settle for the factory patches—they're meant to be adjusted as needed.

Audio example 67: Small hall reverb

Audio example 68: Small room reverb

Audio example 69: Plate reverb

Audio example 70: Reverb with long RT (2.6s)

Audio example 71: Reverb with short RT (1.0s)

Delays

A delay processor simply takes an incoming signal, holds it for a certain amount of time, then outputs it back to the mix. In the really old days we used tape machines. The physical distance between the recording and playback heads provides a built-in delay, so recording a signal to another machine in the room and monitoring off the playback head returns a signal slightly late compared to the original. This was the origin of the "slapback" echo effect; John Lennon's vocals are a great example of this on the Beatles' recordings. Changing machine speeds also varied the delay time, and if you continuously varied the speed other really interesting things happened. This got easier with tape-based delay devices such as the Echoplex, where the box actually had a continuously-running loop of tape inside. Feed it a signal and the output timing could be adjusted by a slider moving the relative head position.

Digital delays input a signal, convert it to digital data if it's analog, store the numbers in a buffer for the desired delay time, then output the result. Sound quality for modern units, including plug-ins, is far better than the old tape-loop systems, but since engineers are still rather fond of tape-based echo, you can find slap delays on tape emulation plug-ins complete with that analog tape sound. Guitar players can get in on the action with a modern Echoplex delay pedal. Really.

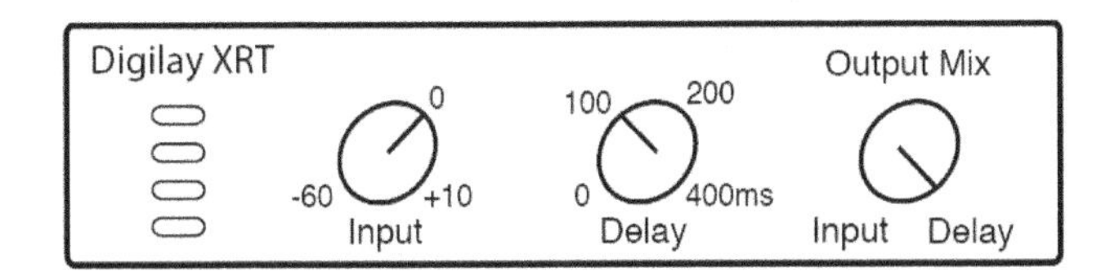

Settings

Delay time

Determines how long the processor will "hold" the original signal before sending it out. Internally it's making a copy of the original, and the user decides how much of each is heard at the output.

. . .

Feedback

Tape slap generates a single instance of an echo. Delay processors provide the capability to loop that echo back to the input of the unit, where it gets delayed in addition to the original signals. This cycle, called regeneration, continues triggering multiple delays.

Modulation

The concept of modulation is to superimpose a signal onto another in order to control some aspect of it. In this case we employ a low frequency oscillator (LFO) to vary the overall wave shape of the audio signal being delayed. This varies the sound of the track in a variety of ways that are best heard than described, but include flanging, chorus, and frequency modulation. Check out the audio examples below for this to make more sense.

Tap

Delay times can be entered manually if you know what you're looking for, but delay processors also provide a button or pad for the user to tap a tempo with your finger or mouse. Play your track and tap to the beat; you'll see the delay time adjust accordingly.

Application

Delay is useful for a number of things. The traditional slap echo is very effective, especially for vocals. Set it for around 60ms and bring the level down a bit; you may not want it calling attention to itself and competing with the main vocal. Another example is running a copy of a single rhythm guitar track (or organ, etc) through a delay, then panning the delay signal different from the original. The dual guitar track, when delayed, provides more depth to the part and the delay helps it move more in the mix. Vocal delays other than slap are employed a great deal, but often in subtle ways you won't even notice. Multiple instances of different delays on a lead vocal might be panned around the main track, but at a low level where you don't really hear them; mute the delays and you'll miss it, turn them back on and they just add a feel to the mix.

Make sure you set the delay times to match the song. This could be quarters, eighths, sixteenths, triplets, or whatever, but make it work musically. Also note that with very short delay times, say less than 15–20ms, if both signals are fairly equal in level and panned together you'll get severe phasing that changes the overall sound of the track. Comb filtering throughout the frequency spectrum occurs because different frequencies have different physical wavelengths, and if the time relationship is offset between two identical signals, the acoustical summing of the waveforms results in some frequency components getting louder, others softer. Experiment with delay times and pan these away from each other to get a rich, subtle motion effect that works quite well in a mix.

Audio example 72: 30ms delay added to right channel

Audio example 73: 60ms delay added to right channel

Audio example 74: 120ms delay added to right channel

Audio example 75: Vocal delay (out, in)

Other effects processing options

Doubling

You can make a track sound fatter by generating another copy of it and delaying it very slightly. You won't hear the separate attack, but the double-attack thickens the sound just a bit. Make sure to play with the delay time, since short delays can wreak havoc on the timbre of your track.

Acoustic doubling

A better way for doubling a track is to have the musician record another take on a separate track; since no two performances will ever be identical this results in a richer, thicker sound. Along with the minor timing differences (delay), this method also provides slight intonation differences which contribute to the effect.

Chorus

Ever thought about what makes a choir sound different from a solo singer? All those voices cannot sing exactly the same—there are slight differences in timing, timbre, and intonation. Aside from using a real choir, you can recreate this effect with an effects processor. The *chorus* setting will take the incoming original signal and make a copy. The copy is delayed and slightly detuned, then combined with the original, resulting in a shimmering, fuller sound. Not quite as good as the real thing, but it helps give a track more interest and life. Chorus is often used for background vocals, keyboard pads, organ, guitars, and so on.

Audio example 76: Chorus effect

Flanging

Flanging occurs when a copy of a signal is continuously varied in time relationship with the original. Run a track through a short delay, then constantly vary the delay time while listening to both signals. The combination produces an ethereal sound quality. The term *flanging* came from the practice of having two identical tracks on two different tape machines—you start both at the same time, press slightly on one machine's reel (flange) with your hand to slow it down, then release to let it speed back up. You're causing an extreme phasing relationship between both signals.

Flanging is very effective for guitars, vocals, drums, etc. It's a more pronounced effect than chorus. Try experimenting with either flanging or chorus to help keyboard patches such as organ come alive with more depth and movement. Send a keyboard track to the processor, bring the flanged output back into a separate track, add verb to both (so they both have similar ambiance), and pan apart. Remember to EQ similarly also. Or maybe not. Use your ears and play around with it.

Audio example 77: Flanging effect

Phasing

Producing a sound very similar to flanging, *phasing* uses very narrow bandwidth filters (remember the EQ chapter?) that are swept up and down the frequency spectrum. Two copies of the signal are used, one with the filter(s) sweeping up and down, the other not affected. When combined, the filters create phase shift which cancels against the original signal. This cancellation affects different frequencies as the filter moves up and down, producing a very similar ethereal effect as flanging. Very popular and common for guitars.

Audio example 78: Reverb with phaser

Time compression/expansion

Sometimes you need to make a 62 second audio track fit into a 60 second slot, such as for radio commercials. DAWs provide the capability to "shrink" an audio file (or make it longer). Only in small amounts, though, or the processing artifacts start taking over (aliens coming home to roost).

TEN
DIGITAL AUDIO

Sound is an analog event, but there are numerous advantages for encoding this as digital data. Analog recording is inherently non-linear and chock full of issues for recording and reproduction (playback). Digital information can be stored, transmitted, and altered with little or no degradation of sound quality. At its core, digital is simply a string of zeros and ones, known as binary. This can be represented as on or off, so the system only needs to be able to distinguish between two values. Incredibly, such complex information as audio signals, metadata, error correction, and so on can be generated by combining strings of these two digits into meaningful digital words. The theory has been around for generations, though it took a while to develop the process and hardware so it sounds as good as, and in some ways better than, analog.

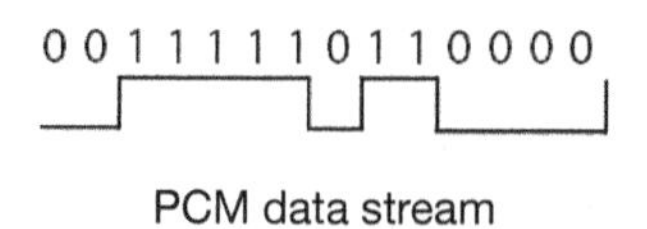

PCM data stream

Sony released the first commercial pulse code modulation recorder in 1977. I'm not quite that old, but while in college in the late 80s we had the improved version of that model, the PCM-F1. And what a system it was! Audio fed into the processor unit was converted into digital PCM data; this was then stored onto a standard Sony Beta video tape recorder. Of course you couldn't grab the tape and run out to the car to check your mix, as it was only playable on the PCM. But we could edit, sort of. The video display (no color, of course) showed an image that looked like a sonar "waterfall" display, where lines or waves of static move across the screen. Some areas of this noise would be tightly bunched together, marking a transient. You couldn't hear a thing, but once you got the hang of it you could tell snare hits and downbeats. We'd set an edit in-point, find the out-point, then pressed record for the system to make the corrected recording to a second VCR. No undo, no audio, no confidence. But it was digital editing.

Over the years the industry developed a variety of digital formats and storage media, including the washing-machine sized DASH tape recorders, tiny DAT videotapes, and of course the ubiquitous compact disc. Some sounded better than others, and generally speaking as time went on digital systems got much better and more appealing to musicians and engineers. In the 1990s the concept of perceptual coding came in the form of Mini-Disc and Digital Compact Cassette, two competing formats designed with consumers in mind. The idea was to reduce data, making it possible to store music on portable media with limited capacity. Perceptual coding attempted to identify aspects of a musical signal that were masked within the mix and therefore inaudible to a listener. It sorta worked, mostly, but the idea was so revolting to the music industry, and confusing to the public, that those formats never caught on. They were also rendered essentially irrelevant with the development of the recordable compact disc (CDR), which let consumers make their own recordings at a quality far superior to the analog cassette.

Today, of course, everybody listens to perceptually coded music, primarily as mp3 or AAC (advanced audio coding). These are known as *lossy* formats, meaning the data compression techniques throw away actual audio information in the pursuit of small file sizes. We're not talking about dynamic range compression, but rather how a computer manages data files. There are also *lossless* formats, such as FLAC (free lossless audio codec) and ALAC (Apple lossless audio codec), that reduce file size as much as possible while preserving original audio information. Naturally a lossy format will give you a much smaller file size, and since network bandwidth and portable device storage was rather limited for years this became the practical, widely adopted solution.

Raw digital audio is usually recorded as WAV (WAVeform audio format) or AIFF (audio interchange file format) files. These utilize no compression whatsoever, capturing the original analog sound as-is. File sizes are large, depending on the resolution used, but everything is there. Engineers always record as high a resolution as feasible, then work downward depending on the ultimate application. If you've ever tried enlarging a low resolution jpg photo, the pixelation jaggies become obvious. Same with audio—always start as high as possible.

So exactly how is an audio signal encoded? We'll simplify it somewhat here, understanding that it's a fairly detailed, complicated process underneath the hood. First, remember that an analog sound is a continually changing event. Compare an analog clock with a digital one, where the second-hand smoothly rotates around the clock face. A digital clock will increment in steps. For a digital recording we have to convert an ever-changing signal into a series of steps. Think of it as a grid, where we plot points on the grid that match the waveform. The grid points are then encoded as data.

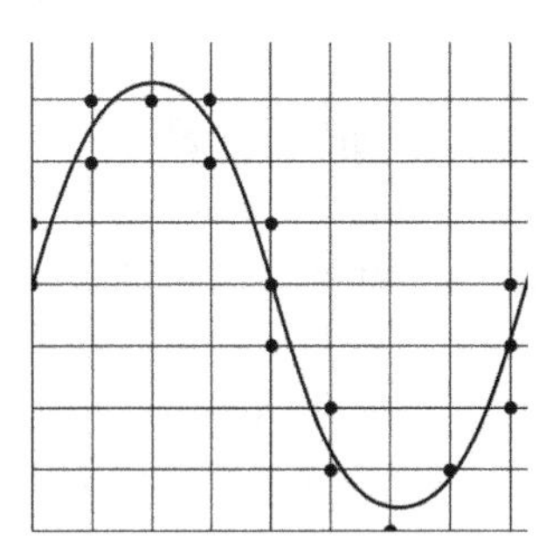

In this example the grid is very low resolution, meaning there are few points of intersection on the vertical (dynamic range) and horizontal (frequency/time) axes. Once the system reconnects these dots during the digital–to–analog conversion for playback it won't look like the original waveform. It's distorted, known as quantization error. But increase the resolution enough and there are sufficient data points to capture the nuances of what we can hear. *Sampling rate* is the number of "snapshots" taken per second of time (the horizontal axis) and *bit depth* represents amplitude (vertical axis).

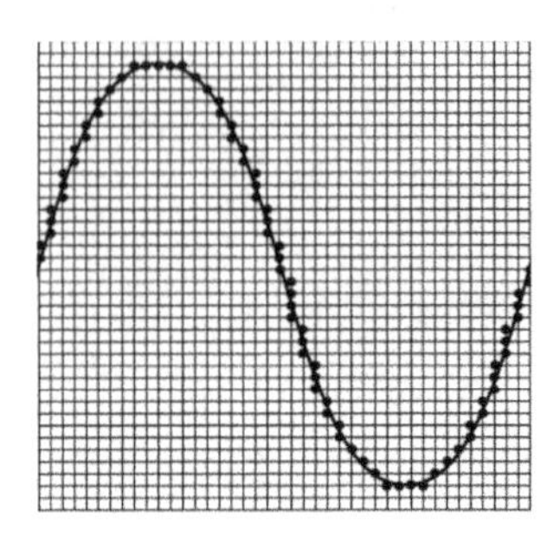

Harry Nyquest was a communications engineer in the first half of the 20th century who helped develop the roots of modern communications theory. Based on his work, the idea is that it requires at least twice as many samples per signal bandwidth to accurately capture it without generating unintended components. What happens is that if an audio signal gets sampled, or encoded, at less than 2x its highest frequency component, extra wave cycles will be introduced into the signal. This is known as aliasing, and so to prevent this the system will do two things: sample the audio at least twice the highest frequency while a low-pass filter removes everything beyond that. You've probably heard the number 44.1kHz associated with digital audio recording. This is the sampling rate specification for the compact disc, representing higher than twice the range of human hearing. The video industry adopted 48kHz for reasons video people understand (or did a long time ago), and improvements in technology have resulted in sample rates as high as 192kHz. Go lower than 44.1, though, and it won't sound very good, especially for music.

Bit rate determines the resolution of dynamic range captured in each sample. Higher bit rates use longer digital words that can store more information. For example compare an 8-bit system (from the 80s) that has a word length of 8 digits. This gives you 256 unique values to encode a signal. It's quite crude and doesn't come close to matching the ever-changing analog waveform. A 16-bit system like the compact disc provides 65,536 values, much better. Current 24-bit processes, though, make the quantum leap to 16,777,216 values. Whereas for 16-bit recording we have to watch levels (don't overload, yet don't lose low-level signals), 24-bit recording gives you so much fine detail and range we can finally re-think old habits. Originating from the days of analog tape, the mindset was to push as close to the top as possible, avoiding distortion, in order to maximize available dynamic range without losing audio in the noise floor. That doesn't apply with 24-bit systems. Keep in mind that good levels are good levels, always, so make sure the meter is showing a healthy signal. For most everyday recording a 24-bit/44.1 or 48kHz setting will work fine. If storage is not an issue go for a 96kHz sample rate (twice as much data).

Most of the magic happens behind the scenes and you don't have to worry about it. A couple of issues, though, to be aware of. *Latency* refers to a delay between the original signal and what comes out of a digital system in real-time. This isn't much of an issue during mixing these days, unless you're running a ton of processor-intensive plug-ins on certain tracks. In a tracking session the problem is feeding a monitor mix during overdubbing. The computer is trying to play back tracks already recorded while inputting new audio to be digitized and routed back out to the headphones. Digitizing and processing audio takes time, even with current computer processing power, and the delay can be audible. Audio interfaces thus have an input/output monitor balance pot so you can dial it to whatever works for each situation. Some interfaces allow you to set up monitoring from the interface itself *before* it hits the DAW, keeping the computer out of the loop and therefore nearly eliminating latency issues.

Clock is the other term that comes up when working with multiple devices. The only way that long string of zeros and ones makes sense to the system is knowing exactly when each word starts. The problem is that this can get skewed, a phenomenon known as *jitter*. Timing is everything, and so a complete digital system much be referenced to a single, solid clock source. Set one device as the master; in some cases a dedicated clocking device must be added into the equipment chain. Sounds crazy, but this actually affects the quality of the audio. Some digital consoles, for example, sound better when clocked with an external, built-for-purpose clocking device.

There are several connection formats for digital audio. Home audio/theater components usually have an optical jack (Toslink) that transmits stereo and compressed multichannel audio (surround) using the S/PDIF format; some devices use the ubiquitous RCA jack instead. HDMI embeds multichannel audio with the video signal. ADAT Lightpipe (Alesis optical protocol), TDIF (Tascam digital interface format), and MADI (multichannel audio digital interface) are examples of uncompressed, multi-channel connections. The groundbreaking trend at the time of this writing is converting digital audio to a format that can be transmitted across standard computer networks. Primarily gaining its footing in the live sound industry, the idea is to put a network jack (CAT5) on each audio device, connect it to a network switch, then use software to route audio data from any device to any other device on the network. This is revolutionary in that no longer do we have to worry about how to split a signal to go to two places. No longer does it take expensive equipment and cabling to connect signals to other devices, locations, etc. Need to set up a remote viewing room for an event that's sold out? No problem—just run an inexpensive CAT5 cable from the network switch to the playback gear in the room and route from the computer. Recording studios with multiple tracking rooms and isolation booths once required complicated cabling and patching systems; now just plug the mic into one of these interfaces and it will come up in the control room. Another example: at my college we often record musical events in the chapel on campus. This building is not connected to where our studios are located, so for decades the only way to do this was to have a complete recording system in the chapel. Now we can plug mics on stage into the campus network and see signals in our control room. My sound system at church runs off a single CAT5 cable from the console to the back room mic preamps and amplifiers. I just rewired our stage with CAT5 jacks in addition to regular XLR mic inputs. The possibilities are endless, and it'll be interesting to see where it all goes.

FINAL THOUGHTS

Still awake? There's lots more audio stuff to discover if you so desire, but that's enough for now. You should have a pretty good idea of how you can improve your show quality, so go practice. Listen to lots of great shows and focus on the *production*—sound quality, timbre of the voices, volume balance between parts, quality of the music, loudness levels, and so on. Use those high-end shows as your target—keep the bar high and don't settle. Remember, *if bad sound were fatal, audio would be the leading cause of death.* Let's save the world, shall we?

Here is a short list of podcasts that excel in audio and production quality. Yes, most of them are produced by a team of professionals, but again, here's your reference point.

- *Radiolab*: Considered one of the best, this show is produced out of the WNYC studios.
- *99% Invisible*: This very high quality, independent production is about "all the thought that goes into the things we don't think about". At over 250 million downloads, it's one of the top shows around.

- *Serial*: A runaway, award-winning hit series that tells a true story over the span of an entire season.
- *Twenty Thousand Hertz*: Excellent audio produced by the audio engineers at Defacto Sound.
- *Themed Attraction Podcast*: Of course I have to mention my own show, which I think can hang with the others here. We may not have the same resources or production teams, but we do okay considering it's an all-volunteer effort, our co-hosts are on the other side of the country from where I live, and it's mainly for fun.

And there are lots more, but you get the idea. At least begin to hear the difference between the enthusiastic enthusiast who has no idea about mic placement, room acoustics, EQ, or even dialog balance, and the super-slick, elegant works of art from the professionals. Your goal is to slide right there in the middle somewhere, so let me know how it's going. I hope this has helped you move the needle just a bit farther.

ABOUT THE AUTHOR

Dr. Barry R. Hill is a professor of audio engineering, instructional designer, and writer. He is Director of the Audio & Music Production degree program at Lebanon Valley College in Pennsylvania and a member of the National Academy of Recording Arts and Sciences (Grammys), Audio Engineering Society, and the Themed Entertainment Association. As a partner of ThemedAttraction.com, he produces a podcast on the themed entertainment industry.

Dr. Hill holds degrees in Instructional Design from The Pennsylvania State University, Music Technology & Interactive Media from New York University, and Music with Recording Arts from the University of North Carolina Asheville. He can be contacted at www.barryrhill.com.

ALSO BY BARRY R. HILL

Recording Audio: Engineering in the Studio

Mixing for God: A Volunteer's Guide to Church Sound

CPSIA information can be obtained
at www.ICGtesting.com
Printed in the USA
BVHW042201280121
599067BV00018B/297